鼎爺廚房

原汁原味功夫菜

3

Grandpa's Kitchen 3

李家鼎（鼎爺），香港著名藝員，自幼受父親培訓，操練出一身好廚藝；同時由於天賦、興趣和不屈不撓的熱血精神，武藝與馬術亦相當精湛，可謂廚武雙全。

於 2016 年冬，他主持烹飪節目《阿爺廚房》，在節目中鼎爺大顯身手，親自下廚，炮製拿手的粵式小菜、湯水及甜品。他的刀工、他的廚藝，讓觀眾認識、欣賞鼎爺的另一面，亦讓觀眾重新認識粵菜的細緻與美好。

於 2017、2018 年夏，他分別出版了兩本食譜《家傳粵式手工菜》和《懷舊風味撚手菜》大受歡迎，多次再版。今年鼎爺再接再厲推出《原汁原味功夫菜》，讓一眾擁躉增添飯桌上的美味佳餚。

鼎爺，用心做好每一道菜。

他的執着，在於對味道的堅持。

他烹調的菜式着重原汁原味，只用基本的調味料帶出食物的鮮味，由選料、
烹調方法、爐具至調味料，都有他的一套哲學和心思。

食譜誕生的緣起

在商討本書的大綱時，談起砂鍋，鼎爺
說喜愛用砂鍋的原因是，砂鍋耐熱、存
氣、能保持食物的香氣等等。砂鍋有這
麼多優點，所以怎能沒有煲仔菜呢？

當大家都傾得口沫橫飛、互給意見時，
鼎爺突然說句：「養育小孩不容易」，
由此再說起舊時產婦的調理方法，他家
傳豬腳薑的炮製方法，煮好後如何如
何，我們單憑幻想也覺很美味，所以怎
能不發掘多些食譜，因此有「家傳坐月
子食譜」和「保小兒安康湯。飯」的章
節誕生。

此外，鼎爺還與大家分享他喜愛的小
炒、蒸餸、湯、飯、小吃等等。看鼎爺
的食譜彷似上了一堂烹飪課，他娓娓道
來每道菜的竅門，讓你從根本學起。

明火與砂鍋好餸

煲仔菜的用料可樸實、可奢華,它那惹人垂涎、熱辣香濃的風味,竟然和明火有好大關係。

明火烹調之好處

鼎爺炒菜講求鑊氣,他喜歡以鑊炒餸,鑊底能完完全全接觸明火,炒餸自然鑊氣十足;至於長時間燜煮,鼎爺則愛用砂鍋,利用火焰強力的明火爐頭煮煲仔菜。由於砂鍋以砂及陶土製造,可保存菜式的香氣,久久不散,而且受熱均勻、熱力耐存,蓋上鍋蓋烹煮後,餘溫令菜式保持熱度,在整個晚餐時段能享受熱辣辣的菜餚。

砂鍋保養技巧

鼎爺教大家一個保養砂鍋的小竅門:新買回來的砂鍋用水浸一晚,翌日再盛水煲滾,然後在室溫待冷卻。水倒後抹乾,塗上油並燒熱,抹去油,讓砂鍋吸收油分,令質地堅韌、耐用、防漏,減少長期使用而出現的裂縫。

鼎爺在書內推薦了幾款以砂鍋烹調的必食菜式,用明火配搭砂鍋作菜,明火,火力強;砂鍋,耐熱傳氣,相輔相成,帶來無窮的滋味美食。

煎焗大白鱔

材　料

白鱔	約 2 斤以上
粗鹽	
生粉	
葱	各適量
紅葱頭	
蒜頭	
薑片	

調　味　料

糖	
鹽	各少許
米酒	

醃　料

胡椒粉	
紅葱頭蓉	
蒜蓉	各少許
薑粒	
米酒	

做　法

1. 葱切段；用刀拍扁蒜頭、紅葱頭。
2. 白鱔用粗鹽捽勻，並放在水喉下一邊沖水一邊用百潔布擦去潺液，洗淨，用布吸乾水分。
3. 將白鱔切成粗段，用醃料醃一會。將白鱔撲上生粉。圖 1~3
4. 熱鑊下油，下白鱔（煎前拍去多餘的生粉），用大火煎至七成熟，盛起備用。圖 4
5. 燒熱砂煲，下油，爆香薑片、蒜頭、紅葱頭，下一半葱段，炒至葱軟身時加入白鱔，加蓋焗約 1-2 分鐘，下糖，加蓋焗約 30 秒，下葱和少許鹽，再加蓋焗約 30 秒，關火，於煲蓋淋上米酒即可。圖 5~7

· 宜挑選二斤以上的白鱔，肉質才爽口彈牙。

· 用粗鹽和百潔布揉去白鱔的潺液，比用熱水好；因爲用熱水去潺，萬一時間控制得不好，會令鱔肉變老。盡量要去清潺液，否則吃時會有食漿糊的感覺。

· 將蒜頭、紅蔥頭拍扁才爆香，會更加出味，令這個鱔煲更香。因爲白鱔是淡水魚，帶點腥和泥味，所以要用醃料辟腥味增香。

· 白鱔帶點焦糖香，竅門是先下糖，糖溶化後黏上鱔身，受熱後會呈微微的焦糖化。

煎焗大白鱔

11

柱侯牛腩

材料

牛坑腩	2 斤
白蘿蔔	1 條
冰糖	3 兩
柱侯醬	3 湯匙
薑片	10 多片

香料

八角	6-7 粒
草果	4 粒
桂皮	2 塊
香葉	8 片
甘草	6 片
白胡椒粒	半兩
陳皮	2 個

做法

1. 白蘿蔔去皮、切滾刀塊。白胡椒粒略拍裂。

2. 牛坑腩切件；鑊內注入凍水，放入牛坑腩、數片薑、冰糖約 1/4 兩、半個陳皮、少許白胡椒粒，不要冚蓋，汆水至牛坑腩釋出血污泡沫。撈起牛坑腩，沖洗乾淨，瀝乾水分。

3. 燒熱鑊下油，將餘下的薑片泡油。

4. 砂鍋內注入水，放入已泡油薑片、香料和冰糖，煮至冰糖溶化後放入牛腩，用中火燜約 1 小時，測試牛腩是否已腍；如已腍，下柱侯醬，轉為中慢火燜一會。

5. 放入白蘿蔔，燜約 20 分鐘至白蘿蔔腍身即成。

- 坑腩是近腰肋骨的肉，因爲除去肋骨後會出現一條條坑，所以稱爲坑腩。

- 坑腩牛味濃郁，少肉臊味，耐燜，最適合用作燜或煲湯。

- 燜牛腩時不宜過早放入柱侯醬，這樣會「喧賓奪主」，因爲這時牛腩尚未出味，濃郁的柱侯醬會蓋過牛腩的味道。

- 宜選用日本大根，因爲水分多、肉質細嫩、沒有渣。

- 經過泡油的薑片，因爲肉裏的水分減少，與牛腩同燜後，吸收了牛肉香和醬汁香，非常美味。

陳皮浸乳鴿

材料

大乳鴿	2 隻
陳皮	2 兩
黃砂糖	少許

做法

1. 乳鴿劏好洗淨；鑊內注入凍水，放入乳鴿，不要冚蓋，汆水至乳鴿釋出血污泡沫，洗乾淨，瀝乾水分備用。圖1

2. 陳皮浸軟，刮去瓤，切成大塊。圖2

3. 砂鍋內，加入水（水以浸過乳鴿為合）、陳皮和少許黃砂糖熬1小時，放入乳鴿，大火煲3分鐘，熄火焗40分鐘即可。圖3~4

・何謂陳皮？經曬製及牧藏五年以上的新會柑皮才可稱爲陳皮，不足五年的，只能稱爲柑皮或果皮。

・爲甚麼不用不鏽鋼煲而用砂鍋呢？因爲砂鍋能儲熱，經過四十分鐘後，水溫仍能保持一定溫度。

陳皮浸乳鴿

順德家鄉燜鮏魚

鮒魚	1 條
火腩	半斤
蜆	半斤
蝦	4 兩
青紅黃西椒	各 1/4 個
生粉	
薑片	
紅葱頭	各適量
蒜頭	
葱段	

糖	1 湯匙
鹽	1 茶匙
生抽	1 湯匙
老抽	少許

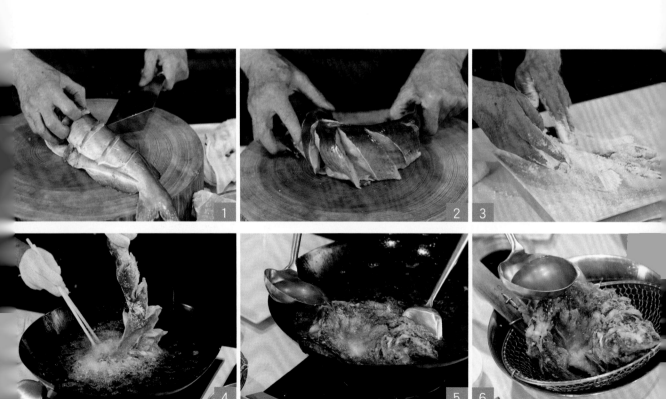

做 法

1. 青紅黃西椒切角；火腩切件；蜆吐砂後洗淨。
2. 鮒魚洗淨、索乾水分後，在一邊魚身輕切數刀，再整條魚撲上薄生粉。圖1~3
3. 熱鑊下油，拍去鮒魚多餘的生粉，將鮒魚彎曲成新月形，炸至五成熟，撈起，瀝乾油。緊記：炸鮒魚時，宜將魚頭先放入油內，因魚頭較難熟；也要不時在魚背澆油，因魚背也較難熟；撈起魚後，也要在魚身補澆一兩杓滾油。圖4~6
4. 燒熱砂煲，下油，爆香薑片、紅葱頭、蒜頭，加入火腩，待火腩的油溢出，倒入蜆，炒勻後加入少許水和葱段，蓋上煲蓋煮至蜆開口（約半分鐘）。
5. 放入蝦（不要蓋鍋），煮至蝦變紅色後，加入青紅黃西椒，撈勻後下糖和鹽調味。盛起部份材料，加入鮒魚，再放入已盛起的材料，下少許老抽調色，蓋上煲蓋，以中大火煮約5分鐘，熄火，下生抽調味，再焗1分鐘即可享用。

鼎爺話你知

· 宜選購頭小、身體又圓又短的鮒魚。
· 一定要等蜆開口後才下蝦，否則蝦肉會煮老；同時煮蝦時不要蓋上鍋蓋，否則蝦肉會變粗韌。

順德家鄉燜鮒魚

黑豆燜塘虱

材 料

塘虱	2 條
黑豆	半斤
陳皮	1/4 個
紅棗	8 粒
蒜苗	1 棵（切段）
紅葱頭蓉	
薑粒	各適量
生粉	
粗鹽	

調 味 料

柱侯醬	1 湯匙

做 法

1. 陳皮浸軟；紅棗沖淨，去核。

2. 用白鑊把黑豆烘至皮裂，見到青肉；加入水，煲至水滾，再用笡箕瀝去黑豆水分。

3. 將適量水、黑豆、陳皮、紅棗放入砂鍋內，煲約 40 分鐘。

4. 在燜黑豆時，用粗鹽擦去塘虱的潺液，沖淨後用乾布索乾水分，撲上生粉炸至半熟。圖1

5. 待黑豆腍，放入塘虱圖2，燜約 10 分鐘，此時塘虱已索滿汁液。下蒜苗段、紅葱頭蓉、薑粒，蓋上鍋蓋燜一會，轉調中火，下柱侯醬，試味，再燜一會即可享用。

· 質素好的黑豆稱為「黑皮青」，豆肉是青色的。

· 因豆總有股豆腥味，所以在烹調前宜用白鑊將黑豆炒至皮裂見青肉，讓豆腥氣味揮發掉；而且經炒過的黑豆口感較鬆化、味道較甘香。

· 街市會有兩種皮色的塘虱出售：灰黑色及黃沙色，宜購買黃沙色的；它的肉質較嫩滑、味道鮮香，沒有泥味。

· 塘虱經撲粉、油炸後，肉質會較結實，燜後能保持完整，而且味道甘香。

· 塘虱炸至半熟就可以，如果炸至全熟，會較難入味。

· 我的小時候，鄉民以此道菜調理身體，因有補血的療效，建議女性食用。

黑豆燜塘虱

山斑魚淮杞燉瘦肉

材　料

山斑魚　　　2 條

瘦肉　　　　半斤

淮山　┐

杞子　├─ 合共 20 元

圓肉　┘

金華火腿　適量

做　法

1. 淮山浸約 1 小時，沖淨；杞子、圓肉略沖洗。

2. 山斑魚劏淨，用薑片汆水。圖 1~2

3. 瘦肉洗淨，切大粒，與火腿一起汆水。

4. 把山斑魚、瘦肉粒、淮山、杞子、圓肉和火腿全部放入大燉盅內。注入
   滾水，封上耐熱保鮮紙或紗紙，燉 3 小時即成。圖 3~6

山斑魚淮杞燉瘦肉

鼎爺話你知

· 山斑魚有滋補強身、補血健脾、增強抵抗力等功效。

· 將山斑魚汆水除了可以去潺外，加了薑片同汆水也能去腥。

· 山斑魚與圓肉、火腿等材料同燉湯，湯水清甜不肥膩。這湯營養豐富，適合開刀分娩的產婦飲用，能幫助內裏傷口癒合。

花生眉豆木瓜大魚湯

材料

大魚魚尾	1 條
花生	8 兩
眉豆	4 兩
青木瓜	1 個
雞腳	8 隻
瘦肉	半斤
薑	4 片

做法

1. 花生、眉豆先用清水浸一晚，沖淨。圖1

2. 魚尾洗淨，瀝乾。熱鑊下油，爆香薑片，下魚尾煎香。

3. 木瓜去皮、去籽、去瓤，切大件。圖2~4

4. 瘦肉切大件，與雞腳一起汆水。沖淨瘦肉和雞腳，斬去雞腳腳趾，並在雞腳肉枕剝一刀。圖5~6

5. 鍋內注入適量水，大火煲滾後下花生，待水滾後煲20分鐘，下眉豆、雞腳、瘦肉，再煲20分鐘，下木瓜，煲約30分鐘，下魚尾，轉用中火煲1小時即可享用。圖7~8

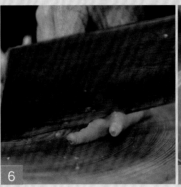

鼎爺話你知

- 這湯有催乳、健脾益胃、滋養皮膚及潤腸通便的功效。

- 處理木瓜時，一定要去瓜瓤，否則湯會有苦味。

- 因為木瓜的質地較腍，要在煲湯中途才加入，如在煲湯初段放入，木瓜會很易溶掉。

- 在雞腳肉枕別一刀的用意，是讓雞腳的骨膠原在煲湯時溢出。

- 飲湯時才在每碗湯下鹽調味，如原鍋湯下鹽，翌日翻熱飲用會有酸味。

花生眉豆木瓜大魚湯

31

豬腳薑

材　料		用　具	
甜醋	5斤	竹墊	1塊
黑糯米醋	4兩	大瓦煲	1個
孖蒸酒	半斤		
老薑	1 1/2斤		
肉薑	3斤		
豬手	2斤		
雞蛋	16個		

家傳坐月子食譜

做 法

1. 薑洗淨，刮去皮，用刀拍扁，放在室外風乾一天或用白鑊烘至水分收乾。
   圖 1~4

2. 請肉檔將豬手斬大件，洗淨後，放入凍水內煮至水滾，將血水污物逼出來。撈起豬手，用凍水洗淨。

3. 煲滾一鍋水，放入豬手煮 10 分鐘，撈起，再沖凍水，瀝乾水分。

4. 瓦煲內放入已洗乾淨的竹墊，注入甜醋，滾後再煲約 20 分鐘，放入薑，轉慢火煲約 40 分鐘，倒入黑糯米醋，煲 1.5 小時後放入豬手，倒入孖蒸酒，用中慢火煲 2 小時，熄火焗一夜。圖 5~7

5. 食薑醋前，以慢火煲滾薑醋，放入已烚熟、剝殼雞蛋圖 8，煮 10 分鐘，熄火浸至入味即可享用。

豬腳薑

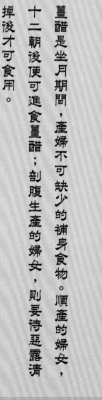 9	10	11

9 10 11

12 13 | 14

15 16 | 17

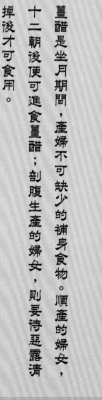

鼎爺話你知

・薑醋是坐月期間，產婦不可缺少的補身食物。順產的婦女，十二朝後便可進食薑醋；剖腹生產的婦女，則要待惡露清掉後才可食用。

・將薑風乾或用白鑊烘乾的作用，是讓薑如海綿般吸滿醋香，薑會比豬手和雞蛋還要美味。

・建議用刮的方法去掉薑皮，因只刮掉表皮，比削皮的更有薑香。圖9～10

・從表面較難分辨哪些是老薑和肉薑（種植時間介於嫩薑和老薑之間），可將薑掰開，如沒有筋、平滑的是肉薑圖11，有筋的是生長期較耐的薑圖12，老薑的筋會較多。

・將豬手汆水兩次，能提升豬手的口感和味道。第一次汆水是將豬手的血水和污物逼出來圖13～14。第二次放入熱水內煮10分鐘，再沖凍水，會令豬皮爽脆、肉質脆。圖15～17

可能各位有疑問：為甚麼會放孖蒸酒？酒可辟去豬手的肉臊味，煲好的薑醋是沒有酒味的，因為已揮發掉。

・瓦煲內放入竹墊，可避免豬手黏鍋。

家傳坐月子食譜

36

田雞焗飯

材 料

田雞	4 隻
白米	適量

（1/3 新米，2/3 舊米）

醃 料

生粉
油
糖 ——┐
紅葱頭蓉 ├— 各少許
薑汁
酒 ——┘

甜 豉 油 汁

生抽 ——┐
老抽 ├— 煮至糖溶
熟油
糖 ——┘

做 法

1. 田雞劏淨，不要脊骨、腳掌，斬件，在田雞腿厚肉位剠幾刀，令其受熱平均。圖1

2. 順序用生粉、油、糖、紅葱頭蓉、薑汁、酒醃田雞，醃約 20 分鐘。

3. 米洗淨，放入瓦煲內，注入水 { 米、水比例是 1：1（水可少一點）}。用大火煲至水滾時，收慢火，待飯面呈蝦眼水，放入田雞，將火調大一點，聽到「嘞嘞」聲和有香味溢出（表示開始有飯焦了），收細火煮 15 分鐘。熄火，撒下葱粒，澆汁料，將飯「翻鬆」，再焗約 5 分鐘就可以享用。圖 2~4

●鼎爺正耳聽、眼看、鼻聞煲仔飯的變化。

田雞焗飯

39

京柿飯

保小兒安康湯。飯

材　料

京柿　　　　3 個
米　　　　　適量

做　法

1. 每個京柿切開三件，備用。

2. 瓦鍋裏放米和水（米、水比例是 1：1），放一半京柿，拌勻，
   加蓋，用大火煮至開始收水時，放入另一半京柿，收細火，
   加蓋（這時要留心瓦鍋裏的聲音，會有「嘞嘞」聲）及有飯
   香後，熄火焗 5 分鐘即可進食。圖 1~4

京柿飯

4

鼎爺話你知

· 這飯有健脾開胃的功效。

· 京柿在中藥店有售。它表面白色的粉末是柿霜，不用洗去，性涼味甘，有清熱、潤燥、化痰的作用。

保小兒安康湯。飯

生熟薏米淡竹葉西施骨湯

保小兒安康湯。飯

材 料

生熟薏米　各 1 兩
淡竹葉　　1 紮
西施骨　　1 斤
燈芯草　　半兩
粟米　　　1 條

做 法

1. 西施骨汆水，沖淨。
2. 生熟薏米浸約 1 小時，沖淨。淡竹葉、燈芯草沖淨。粟米沖淨，切大件。
3. 鍋內加入適量清水，把所有材料冷水下鍋，先用大火煲滾 20 分鐘，改用中火煮 1.5 小時即可。

❶ 熟薏米 ❷ 燈芯草 ❸ 淡竹葉 ❹ 薏米

●粟米

· 這道湯有淡淡的竹葉清香，
與一般肉湯有很大的分別。

· 以前的鄉間幼兒，如有熱
氣、尿黃，父母會煲這湯給
他們飲用。

· 將粟米切大件，可加快散出
香味。

生熟薏米淡竹葉西施骨湯

45

勝瓜蒸魚片

材料

勝瓜	1 條
斑腩	1 斤半
蒜蓉	約 2 湯匙
葱粒	⎱ 各少許
紅辣椒粒	⎰
滾油	⎱ 各適量
豉油	⎰

醃料

生油	
糖	⎱ 各少許
鹽	⎰
生粉	

做法

1. 斑腩洗淨，切片，下少許油拌勻，再下糖、鹽和生粉，拌勻備用。圖 1

2. 刨去勝瓜的稜角，放在水喉下一邊沖水一邊用小刀刮去老皮；然後將勝瓜切成斜片。圖 2

3. 蒸碟鋪上勝瓜片，在每片勝瓜上鋪上魚片，灑上蒜蓉，蒸約 8 分鐘。圖 3~4

4. 灑上葱粒、紅辣椒粒，潷滾油，碟邊淋豉油，趁熱享用。圖 5~6

勝瓜蒸魚片

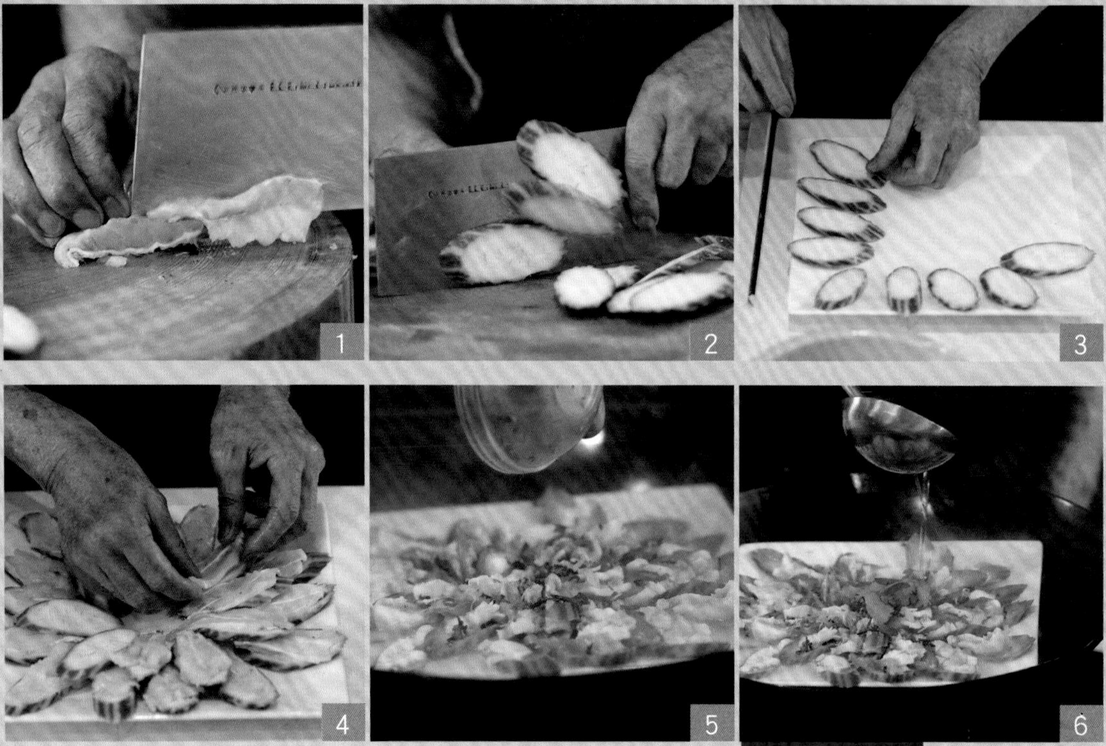

鼎爺話你知

- 勝瓜的稜角很硬，烹調前一定要刨去。因爲老瓜皮會澀口，故要刮去表面的皮青。
- 切魚片前，要先用手摸遍斑腩，感覺是否有魚骨；如有，就要用鉗子拔去。
- 醃魚片時加入生粉，可避免菜式蒸後「水汪汪」。

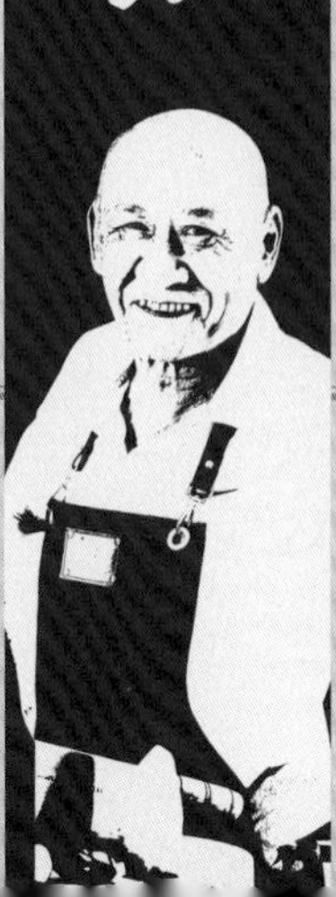

百花豆卜

材　料

豆卜　　　　12 個
蝦　　　　　半斤
肥豬肉　　　1 兩（切幼粒）
中國檸檬　　1 個
糖 ┐
鹽 │
生粉 ├ 各少許
水 │
油 ┘

做　法

1. 蝦去殼、去腸，洗淨後，索乾水分，用毛巾捲好，放入雪櫃冷藏約 1 小時。
2. 蝦肉用刀拍扁或切粒後，用刀背剁碎，順時針方向攪打至起膠，先加入糖，攪勻後順序加入鹽、肥豬肉、水、生粉和油，每次都要充份攪勻。
3. 在豆卜的 1/3 處切開，底面反轉，將蝦膠釀在豆卜內。圖 1~2
4. 燒熱油，要注意油溫不要太高，把釀有蝦膠的一面朝下，慢慢放入油鑊中，炸至微黃色，再轉大火逼出油分，撈起，瀝去油分。圖 3~6
5. 享用時，可擠些檸檬汁在百花豆卜上。

小菜

1 2

3

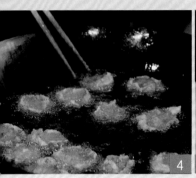

4

5

6

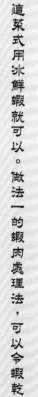

鼎爺話你知

- 這菜式用冰鮮蝦就可以。做法一的蝦肉處理法，可以令令乾身、吃時較爽口。

- 用刀背剁蝦肉，不會將蝦肉剁得太爛，仍可吃到粒粒爽口的蝦肉；否則蝦肉剁得太綿，會沒有口感。

- 蝦膠肉放入肥豬肉，會令蝦膠味道更香，口感軟滑些，吃時不會「硬實實」。

- 蝦膠肉不要放馬蹄粒，因馬蹄肉含有水分，會令炸後的蝦膠質感鬆散。

- 想弄些新花樣？蝦膠釀入豆卜後，可用剪刀在豆卜邊剪幾刀，炸後形成花瓣狀。圖示

- 將豆卜從熱油撈上來時不要熄火，否則豆卜會吸滿油。

百花豆卜

51

自家製XO醬

珧柱	4 兩
蝦米	4 兩
金華火腿	半兩（切幼粒）
紅葱頭	6 粒
蒜頭	2 個
紅辣椒	7 隻
糖 老抽	各少許

●珧柱

●蝦米

做 法

1. 珧柱浸軟，撕去硬枕，將珧柱撕成幼絲。蝦米浸軟，剁碎。

2. 將紅葱頭、蒜頭切成蓉；紅辣椒切粒。

3. 燒熱油鑊，爆香蒜蓉，倒入紅葱頭蓉，炒勻後加入火腿粒，再下蝦米碎，兜勻後倒入珧柱絲，用鑊鏟推散珧柱絲（否則珧柱絲結成一團），下少許水，用大火炒至水分收乾，溢出香味，下少許糖調味（這時要試味），下老抽調色，炒勻後，倒入紅辣椒粒，炒至紅辣椒粒變色後即成。圖 1~7

自家製 XO 醬

53

鼎爺話你知

・切紅辣椒時建議戴手套，避免手指有灼痛的感覺。

・火腿買回來後要試味，這樣煮有火腿的菜式時更能掌握調味料的份量。

自家製XO醬爆蝦球

材 料		**蝦 醃 料**	
大蝦	12 隻	糖	
自家製 XO 醬	適量	油	
京葱	1 棵	生粉	各少許
三色椒	各 1/4 個	鹽	
生粉水	適量		

做 法

1. 大蝦去殼，用布索乾水分，在蝦中間剖開，去腸，在蝦的左右側各剠一刀，加入糖、油、生粉、鹽醃一會。圖 1~4

2. 三色椒切塊；京葱斜切片，每片約 2 厘米厚。圖 5~6

3. 熱鑊下油，下 XO 醬和一半京葱，下少許糖，待京葱變色後，倒入蝦兜勻，下少許鹽，加入餘下的京葱，倒入三色椒，下生粉水，兜勻，盛起，趁熱享用。圖 7~10

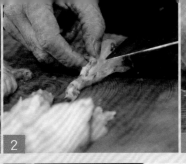

鼎爺話你知

- 京葱分兩次放入，第一次是取其香，第二次取其脆口。

- XO醬的份量視乎吃辣程度而定，悉隨尊便。

自家製XO醬爆蝦球

黃沙蜆蒸水蛋

材　料

雞蛋	3 個
黃沙蜆	12 隻
芫茜 葱花	├ 各少許
熟油 生抽	├ 各適量

做　法

1. 黃沙蜆吐沙後，起肉。圖1~2

2. 雞蛋磕開放入大碗內，加入蜆肉，拂勻，注入水（雞蛋和水的比例是 1:1.5），再拂勻，倒入蒸碟內，撇去蛋面的泡沫，封上耐熱保鮮紙。

3. 用大火燒滾水，將蛋蒸約 10 分鐘，取出。撒上芫茜、葱花，澆上熟油 和生抽，趁熱享用。

鼎爺話你知

- 令蜆吐沙的方法好簡單，除了用鹽外，還可以將約攝氏六十度的滾水倒在蜆身上，這可讓蜆快點吐沙。

- 蛋和蜆倒入蒸碟後，要用筷子分散可能會疊在一起的蜆肉。

- 如不撇去蛋面的泡沫，蛋蒸後會不夠平滑。

黃沙蜆蒸水蛋

豆瓣醬爆蟶子

材料

蟶子皇	1 斤
豆瓣醬	1 湯匙
三色椒	各 1/4 個
京葱	1 棵
蠶豆	適量

調味料

糖	少許
鹽	少許

芡汁

生粉水	2 茶匙

做法

1. 京葱切斜片，三色椒切成角形；蠶豆壓碎，略剁備用。圖 1~3

2. 用刀從蟶子中間劏開，洗淨腸臟及穢物。圖 4~6

3. 燒滾水，放入蟶子略灼，見蟶子肉收縮、離殼，馬上撈起，放入冰水降溫。撈起，用毛巾索乾蟶子肉水分。圖 7~8

4. 燒熱油鑊，倒入一半京葱，炒至京葱變色，呈略透明。加入豆瓣醬、蠶豆，再下三色椒，兜勻後倒入餘下的京葱，下蟶子，炒勻後下糖及鹽，倒入生粉水埋芡，兜勻即可上碟。

小菜

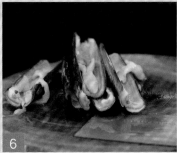
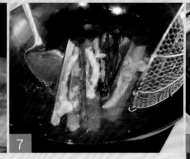

鼎爺話你知

- 新鮮蠶豆的份量視乎口味而定。加入蠶豆能增加香味和口感,因爲現成的豆瓣醬多將豆瓣磨得很幼細,較少內有星粒狀的豆瓣。

- 許多朋友只食京蔥白色的部分,其實青色部分蔥味濃郁,不要棄去。將京蔥切斜片,在炒時多能保持原狀,令香氣得以保留。

- 一定要切去三色椒的瓤,防止三色椒出水。

豆瓣醬爆蟶子

手切鯪魚滑釀豆角

材 料

鯪魚	6 條
青豆角	10 條
冬菇	2 朵（浸軟、切幼粒）
蝦米	1 兩（浸軟、切幼粒）
半肥瘦豬肉	1 兩（切幼粒）

焯 豆 角 料

糖
熟油 ——— 各少許

調 味 料

鹽
生粉
糖 ——— 各少許
熟油

芡 汁

生粉水
蠔油 ——— 各少許

做 法

1. 起鯪魚脊肉，取三份一鯪魚脊切成薄片，略為剁碎。一份用刀刮起肉；
   另一份連皮切片後剁蓉。圖 1~3

2. 在大碗內混合鯪魚肉、半肥瘦豬肉、冬菇、蝦米，加入少許鹽、生粉、
   糖，拌勻後加入少許熟油（如太結實，可加入少許水），撻至有黏性、
   不鬆散就可以。圖 4~5

3. 豆角去頭，放入滾水內灼軟，待豆角變色後下糖，豆角開始變軟後下熟
   油，豆角軟後，撈起待涼。圖 6~7

4. 將豆角結成環狀，釀入鯪魚滑。圖 8~15

5. 平底鑊下適量油，將鯪魚滑釀豆角煎香，放在碟上。圖 16

6. 將蠔油加入生粉水煮成芡汁，淋在煎好的豆角上，趁熱享用。

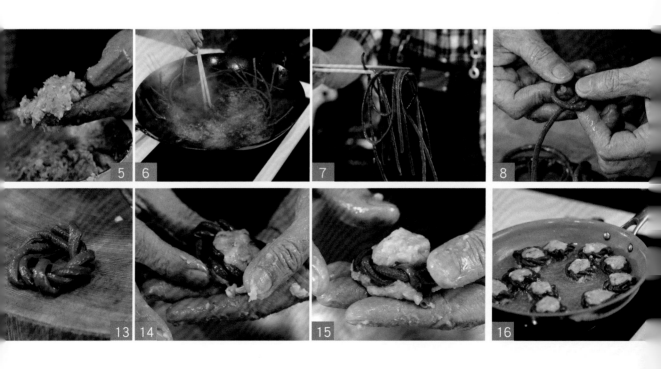

鼎爺話你知

- 這道菜內的鯪魚滑有三種不同的切法，有三種不同的口感，想知道箇中滋味如何，你一定要自己試做。

- 可以用刀或鐵匙刮鯪魚肉。一定要由尾向頭刮去，魚肉就會刮出來；如由相反方向刮，魚骨會刮出來，吃時會鞭骨。圖示

- 宜選幼身的青豆角，而且大小要平均，才能編織美麗的豆角環。

- 灼青豆角時下糖，可去除草青味和不會轉色；下油則令青豆角保持光澤和翠綠。

●魚骨留在魚皮上

五色菊花

材料

鯇魚脊	1 大條
生粉	適量

醃料

鹽	┐
糖	┘ 各少許

汁料

新鮮橙汁	約半杯
檸檬汁	2 湯匙
黃砂糖	1 湯匙
青豆	┐
甘筍幼粒	┘ 各少許

裝飾

松子（烘香）	┐
橙肉粒	│ 各少許
三色椒幼粒	┘

小菜

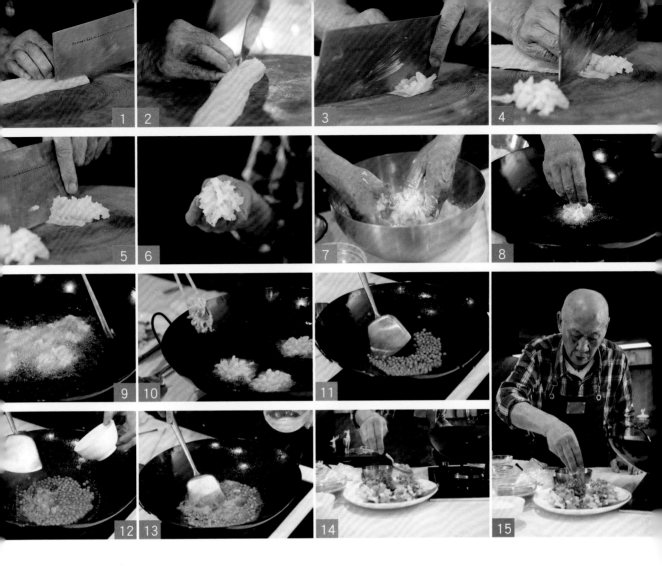

做 | 法

1. 鯇魚脊洗淨，用布索乾水分。魚肉以拖刀手法切出條紋，轉方向切出格
   仔紋，再用斜刀（即是切時要稍微斜斜的往內切）將魚肉切成四方；用
   鹽和糖醃魚肉一會，並撲上適量生粉備用。圖 1~7

2. 熱鑊下油，拍去魚肉多餘的生粉，逐件魚肉放入鑊中炸熟成菊花狀，撈
   起，瀝去油分，放在碟上。圖 8~10

3. 用另一鑊，鑊內下少許水，倒入青豆，煮一會後下甘筍幼粒，注入橙汁，
   加入檸檬汁，下黃砂糖調味，下少許生粉水煮稠汁料。圖 11~13

4. 汁料澆在菊花魚球上，飾上三色椒粒、橙肉粒和松子，即成美麗的五色
   菊花。圖 14~15

五色菊花

一絲不苟的鼎爺

鼎爺事事親力親為，一絲不苟，這個功夫菜上桌不消一分鐘已給吃掉，但預備功夫「多籮籮」。單是製作汁料、用作配襯的食材，鼎爺也認真對待。

鼎爺話你知

· 可以用花尾龍躉、大桂花魚代替鯇魚脊，但價錢略貴。

· 將魚肉用斜刀切成四方的原因，是想呈現上闊下窄的形狀，炸時魚肉會捲起成菊花狀。

梅子瘦肉蒸筍殼

材 料

筍殼魚	1 條（1 斤以上）
酸梅	3 粒
冬菇	2 朵
柳梅	1 兩
鹹酸菜絲	少許
糖	
熟油	各適量

做 法

1. 冬菇浸軟後切絲；柳梅切絲後汆水，備用。

2. 酸梅去核，剁成蓉，與冬菇絲、柳梅絲和鹹酸菜絲拌勻，下糖拌勻，加入少許熟油拌勻後，灑入少許鹽。

3. 筍殼魚劏淨，瀝乾水分，劏開，切去魚骨。 圖1

4. 蒸碟頭尾各放一隻竹筷子，放上筍殼魚（掰開魚腹，讓其趴在竹筷子上），放上冬菇絲、柳梅絲、鹹酸菜絲和酸梅蓉，大火蒸約 9 分鐘；澆下熟油享用。圖2

鼎爺話你知

- 筍殼魚肉質細緻，味道鮮美，沒有土腥味，宜清蒸或油浸。

- 因為蒸魚時間較短，為確保柳梅熟透，蒸前要先汆水。

- 將魚去骨並趴在竹筷子上蒸，讓魚能受熱均勻。

梅子瘦肉蒸筍殼

桂花炒珧柱

材料

蛋黃	4 個
珧柱	10 粒
紅花蟹	1 隻
新鮮竹筍	2 個
銀芽	1 兩
荷蘭豆	1 兩
冬菇	3 朵
糖 鹽	各少許

做法

1. 珧柱浸軟拆絲，一半珧柱絲蘸生粉炸至香脆。

2. 冬菇浸軟切絲，荷蘭豆撕去硬邊、切絲；竹筍切絲後汆水。

3. 花蟹蒸熟後拆肉。

4. 銀芽汆水，瀝乾水分。

5. 蛋黃拂勻。

6. 熱鑊下油，下蟹肉，炒勻後下浸泡過的珧柱絲，推散珧柱絲，加入冬菇絲、筍絲和荷蘭豆絲，下糖、鹽，炒勻後下蛋黃，用中大火不停推勻直至起粒成桂花狀，下芽菜和已搓碎的炸珧柱，推勻，趁熱享用。圖 1~3

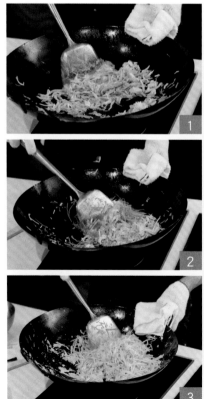

· 這道菜是一個考牌之作，最考炒功，因為火太大，蛋會焦；如太小，就會變成一團漿。

· 建議用湖北出產的紅黃蛋，炒出來色澤金黃。

· 因為銀芽會出水，在炒前一定要汆水，注意時間不要太耐，免得過軟，失去爽脆的口感。

· 拆蟹肉是有竅門的，看看敔怎樣處理吧。

· 新鮮竹筍切幼絲後要汆水，去除苦澀味。也看看敔怎樣處理新鮮竹筍。

豉蒜南瓜燜豬肋條

材　料

豬肋條	1 磅
南瓜	1 斤半
豆豉	約 1 湯匙
蒜頭	
紅葱頭	各適量
老抽	

醃　料

糖	
油	
鹽	各少許
生粉	

做　法

1. 南瓜洗淨，連皮切件，蒸約 25 分鐘，備用。

2. 豬肋條洗淨，瀝乾，順次序用少許糖、油、鹽、生粉醃一會。

3. 豆豉切碎；蒜頭及紅葱頭拍扁，略切。

4. 熱鑊下油，下豬肋條兜炒，下少許糖，兜炒一會後下紅葱頭和蒜頭，炒勻後下豆豉炒勻，加入水，下少許老抽調味，蓋上鑊蓋燜約 15 分鐘。圖 1~3

5. 放入南瓜，下少許鹽，以慢火兜炒南瓜和豬肋條至水分收乾，即可享用。圖 4~5

豉蒜南瓜燜豬肋條

· 南瓜以重手的為佳。我喜歡用中國南瓜，肉質比較綿密，水分較少，煮時不會水汪汪，吃時口感細緻、更滑。

· 南瓜連皮煮，雖然皮較硬，但瓜味香、營養好。

· 豆豉不要切得太碎，否則煮後的南瓜顏色黑黝黝，外觀不美。

· 豬肋條肉是肋骨之間的肉，又嫩又有豬肉香；建議用黑毛豬的肋條肉，肉質更加嫩滑。

小菜

82

鼎式煎羊排

材料

羊架	1 個
孜然粉	2 湯匙
五香粉	1 茶匙
鹽	半茶匙
油	1 湯匙
蒜頭	1 個
紅葱頭	3 粒
紅指天椒	2 隻
黃西椒	1/4 個

做法

1. 羊架抹乾水分，切件，用孜然粉和五香粉拌勻羊排，下鹽，撈勻後再下油，醃約 10 分鐘。圖 1~2

2. 蒜頭、紅葱頭、紅指天椒、黃西椒切粒。

3. 熱鑊下油，先下蒜粒炸至微黃，然後順次序下紅葱頭粒、紅指天椒粒和黃西椒粒，撈起，待用。圖 3~4

4. 燒熱煎鑊，下油，煎香羊排，上碟。圖 5~6

5. 放炸蒜頭紅椒粒在羊排上，趁熱享用。

鼎爺話你知

· 建議選用紐西蘭出口的羊架，肉質嫩滑。今次食譜選用的羊架，能切成九件羊排。

· 在醃羊排時下少許油，令煎後的羊排帶點脆口。

· 所有醃料放進羊排後，宜逐件羊排捏一會，讓它快些入味。

· 用猛火熱油煎羊排，能迅速封着肉汁，令羊肉不會乾韌。

鼎式煎羊排

雞油莧菜

材料

莧菜	1 斤
蒜蓉	1 湯匙
雞油	適量
紅辣椒絲	1 隻份量

調味料

糖	⎱ 各少許
鹽	⎰

煎雞油

雞肥膏	⎱ 各適量
紅葱頭碎	⎰

做法

1. 雞肥膏洗淨，瀝乾備用。
2. 熱鑊下少許油，放入雞肥膏，煎一會後下少許水，讓雞油快點溢出。加入紅葱頭碎，待葱香味釋出後，將油過濾，備用。
3. 莧菜洗淨後，瀝乾備用。
4. 熱鑊下雞油，倒入莧菜，炒約四成熟時，下蒜蓉、糖兜炒，下鹽炒勻，加入紅辣椒絲，炒勻後即可上碟。圖1~3

◎莧菜

鼎爺話你知

· 雞油比花生油、粟米油等更香滑。
· 蒜蓉非常易燶，所以炒菜至四成熟時才下蒜蓉。

雞油莧菜

小菜

冬瓜寶盒

材料

冬瓜　　　一圈約 3 吋寬
金華火腿　4 兩

芡汁

鮑汁　┐
生粉水　├─各適量
糖　　┘
杞子　　　少許
蛋白　　　半個

做法

1. 冬瓜去皮，不要見到皮青；將冬瓜修成長方形，再切成每片厚 1 厘米、長 7.5 厘米、濶 5.5 厘米的長方塊，在冬瓜片中間切一刀，但不要切斷。圖 1~5

2. 金華火腿先蒸 45 分鐘至 1 小時，按冬瓜塊的數量切成薄片（要比冬瓜細少許）火腿釀進冬瓜內。圖 6~7

3. 將火腿冬瓜片蒸至冬瓜透明，約 10 分鐘，視乎厚薄大小而定。

4. 生粉水和鮑汁調勻（煮芡汁前先試鮑汁的味道，如有點鹹可下糖調味。），煮成芡汁，加入杞子和蛋白，推勻，淋上冬瓜面即成。

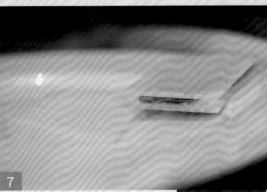

鼎爺話你知

· 切冬瓜皮時要切深一點，因皮青的味道有點澀。

· 火腿要試味，因為質素良莠不齊；如發覺火腿過鹹，可以蒸約四十五分鐘至一小時，以減少鹹味。

· 想用這個菜宴客又想省功夫，可試試簡化版：將冬瓜切薄片，用熱水拖軟，夾火腿後再蒸。

冬瓜寶盒

小炒王

材　料

韭菜花	1 紮
蝦乾	1 兩
銀魚乾	1 兩
魷魚乾	1 隻
鮮魷（細）	2 隻
鮮蝦仁	半斤
菜脯	3 條（可隨意）
紅西椒絲	適量

調　味　料

糖	1 茶匙
鹽	半茶匙

做　法

1. 魷魚乾浸軟後切絲，鮮魷劏淨後剔花，菜脯切絲。圖 1~4

2. 韭菜花切去近根部較老的部分，切段，將頭段、中段和花蕾分開放。

3. 蝦乾浸軟，撈起瀝乾水分，吹至乾透。熱鑊下油，放入蝦乾炸香。

4. 銀魚乾浸透，撈起瀝乾水分，吹至乾透。熱鑊下油，放入銀魚乾炸香，灒少許水令銀魚乾鬆化，一直炸至銀魚乾微黃，盛起，瀝去油。

5. 熱鑊下油，爆香蝦乾，倒入銀魚乾炒勻，放入魷魚乾，兜勻後放入鮮魷、蝦仁，待蝦仁將熟時下糖，炒勻後倒入菜脯絲，加入韭菜花頭段，開始轉色後加入中段韭菜花，炒勻，放入花蕾段，下鹽，兜勻上碟。圖 5~6

6. 最後可用紅西椒絲裝飾。

鼎爺話你知

· 臨上碟前，可灑入已炸脆的松子、花生或腰果，脆口好吃。

· 糖可以令韭菜花保時翠絲。

· 為甚麼要分三個階段放入韭菜花？因為頭中尾段的受火程度各有不同。

· 不要下蒜頭，因為會搶去韭菜的味道。

魷魚乾

韭菜花

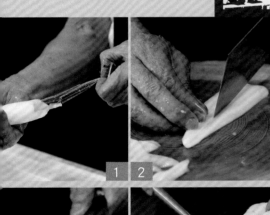

1 2 3

4 5 6

傳統柚皮

材料

厚身柚皮	4 大件
鯪魚骨	1 斤半
大地魚	2 條
蝦米	2 兩
豬油	適量
糖	2 湯匙
鹽	1 茶匙

做法

1. 用火將柚皮燒至焦黑，浸約 7-8 小時，甚至一晚，焦皮開始剝落，代表已浸發完成，可以擦去焦黑部分。洗淨柚皮，擠乾水分，撕去硬筋，裁成橢圓形，然後啤水兩天（期間要榨水 7-8 次，以便去除苦澀味），直至柚皮顏色潔白為止。圖 1~6

2. 將豬油倒入鍋內，放入柚皮，豬油要蓋過柚皮，讓柚皮吸滿豬油。

3. 大地魚放在明火上燒至香味溢出，刮去燒焦的部分及切去魚頭，將大地魚切塊。圖 7~10

4. 煲滾一鍋水，下糖 2 湯匙和鹽 1 茶匙，放入鯪魚骨，倒入蝦米和大地魚，用大火煲半小時，待蝦米和鯪魚骨完全煮出味道後，放入柚皮，並放上竹撻（防止柚皮翻動），滾起後不要蓋緊鍋蓋、要側放，留些空隙；轉中火煲一會，再轉中慢火煲 3 小時。圖 11~12

5. 舀起柚皮放在碟上，煮滾適量原湯，澆在柚皮上，趁熱享用。圖 13~14

小菜

鼎爺話你知

・柚皮吸滿豬油後，吃時入口即溶、軟滑。要怎樣才知道柚皮已吸滿豬油？用瓷羹壓一壓柚皮，如沒有泡沫冒出，代表已吸滿豬油。圖一

・自製豬油非常簡單。在肉檔購買肥豬肉，清洗乾淨，瀝乾水後，放入鑊內並加入少量水以慢火煮出豬油。

・柚皮一定要燒得夠焦黑，才能除去苦澀味。

・用鯪魚煲湯味道鮮甜，如用其他魚如大魚，則帶有腥味。

・煲魚湯時要撇去泡沫，否則湯會苦。圖二

●圖一

●圖二

傳統柚皮

陳皮八寶鴨

材料

米鴨	1 隻（約 2 斤半）
陳皮	4 個
生薏米	1 兩
粘米	1 兩
火鴨	1/4 隻
鮮蝦仁	2 兩
冬菇	半兩
鮮栗子	3 兩
鮮蓮子	2 兩
合桃	1 兩
炸瑤柱絲	1 兩
（做法參考 P.76 桂花炒瑤柱）	
鹹蛋黃	4 個
紗紙	4 張

上色料

滴珠油	適量

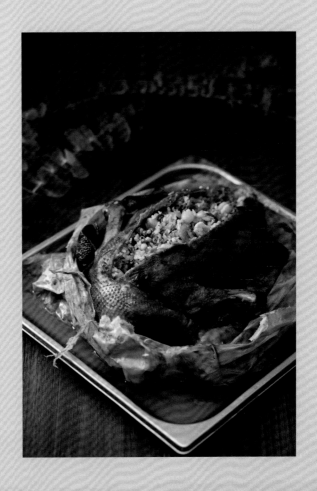

做 法

1. 陳皮浸軟、刮去瓤，取 2、3 片陳皮切絲。

2. 粘米、生薏米浸透，瀝乾水分。鮮蓮子去芯；火鴨、冬菇（先浸軟）切粒；鮮栗子、合桃切半；鹹蛋黃一開四；蝦仁片去蝦衣；全部材料與陳皮絲拌勻成餡料。

3. 切去米鴨尾部兩粒性腺，斬去鴨掌。從米鴨尾部拆骨，用小刀輔助，扯出鴨胸骨。將餡料釀入鴨腔內（約六成滿），用鵝尾針縫好收口。圖 1~10

4. 鑊內注入凍水，放入鴨氽水，不要加蓋，期間要舀滾水至鴨身。撈起鴨，稍涼後（以不燙手為合），掃上滴珠油，在鴨身上疏針。圖 11~12

5. 燒熱油，放入鴨拉油至皮收緊，期間要用湯杓舀滾油至鴨身，撈起鴨瀝乾油。圖 13~15

6. 紗紙鋪上部分陳皮，放上鴨，鴨身再鋪上陳皮，包好，紗紙面再放一片陳皮，放入蒸籠蒸 3 小時即成。圖 16~21

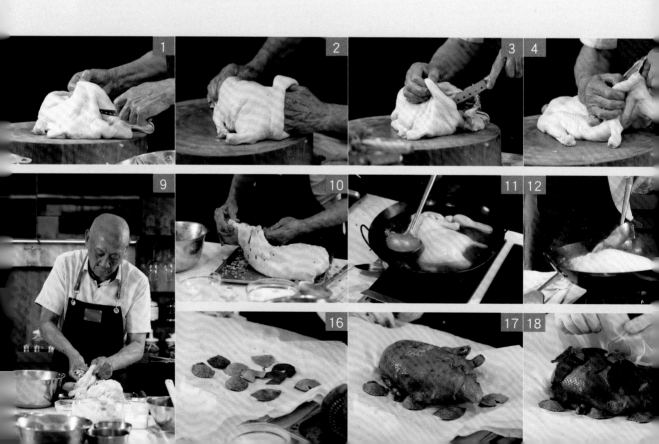

鼎爺話你知

- 米鴨尾部兩粒性腺非常腥，一定要切去。

- 將鴨汆水除了可以去除脂肪、腥味外，還可以收緊鴨皮，容易上色。

- 疏針的作用是讓鴨多餘的水分溢出，也讓不好的味道散出，並讓陳皮的香氣滲入鴨肉。

- 在夏天炮製這道菜，可以用鳳眼果代替栗子。

- 餡料釀入鴨內六成滿就可以，因餡料是生料，蒸後會發脹。

桑葉冬菇蟲草花蒸雞

材料

雞項	1 隻（約 2 斤）
新鮮桑葉	約 20-25 片
金針	半兩
雲耳	半兩
冬菇	3 朵
蟲草花	半兩

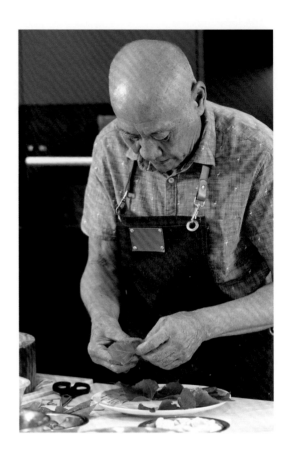

醃料

糖	
鹽	各少許
生粉	
熟油	

做法

1. 金針、雲耳、冬菇、蟲草花分別用水浸軟；摘去金針的硬梗；雲耳去蒂部木屑；冬菇切去菇柄，切絲。圖 1~5

2. 雞起肉切薄片。雞肉用少許糖調味，與金針、冬菇、雲耳、蟲草花一起撈勻，加入鹽、生粉、熟油醃一會。

3. 桑葉洗淨，索乾水分，去葉柄。圖 6

4. 將桑葉鋪在蒸碟上，每片桑葉放上雞肉，並排放其他材料在雞肉上。圖 7~10

5. 用大火燒滾水，將雞蒸約 10 分鐘，熄火焗一會，即可享用。

1 2 3 4
5 6 7
8 9 10

- 桑葉有清肺潤燥、潤腸通便、改善便秘、緩解體內毒素堆積的功效。桑葉未蒸前味道苦澀，口感較粗「鞋」，蒸後則口感嫩滑，皆因蒸時有雞油的滋潤。建議選用較嫩的桑葉。

- 桑葉由採摘至烹調時相隔一段時間，建議浸凍水以保持新鮮和挺身。

- 因桑葉不耐火，雞肉切得越薄越好。

- 雞蒸好後熄火焗一會，是讓雞肉索回蒸汁，雞肉會更多汁和美味。

- 宜用俗稱老鼠耳的小雲耳同蒸，易熟及爽口。

- 蒸煮時間要依雞肉厚薄、大小作出調整。

桑葉冬菇蟲草花蒸雞

玫瑰豉油雞

材料

雞	1 隻（約 3 斤）
冰糖	2 兩
老抽	1 杯
玫瑰露	半杯
鹹水草	數條

滷水料

陳皮	1 個
八角	3 粒
桂皮	4 片
香葉	8 片
甘草	5 片
草果	3 粒

蘸料

薑蓉	
薑葱蓉	各下少許鹽，
紅葱頭蓉	淋下滾油
紅葱頭薑蓉	

做法

1. 雞洗淨，用鹹水草綁實兩隻雞翼，拿着鹹水草將雞放入滾水內燙一燙，並舀滾水進內腔。圖 1~2

2. 煲中注入水（水以浸過雞為合），放入滷水料煮滾，煮至出味後加入冰糖，待冰糖溶化後，下老抽，放入雞（先放雞背）煮約 8-10 分鐘後，加蓋，熄火，浸約 25 分鐘。加入玫瑰露再浸 5 分鐘。圖 3~5

3. 將雞取出，吊起風乾 20 分鐘；雞涼後斬件上碟，伴蘸料享用。圖 6~9

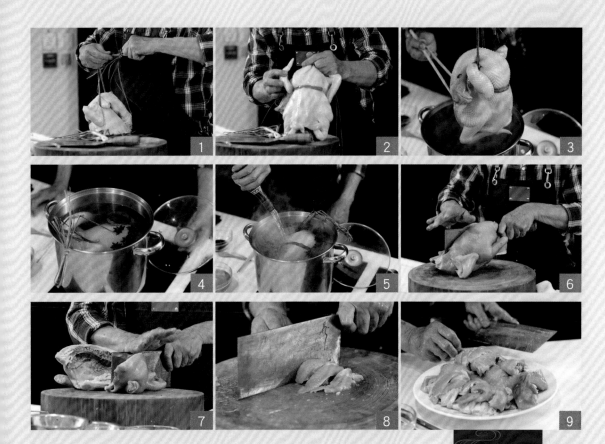

鼎爺話你知

- 做法一是重要的步驟，可以收緊雞皮，防止雞皮在浸雞時爆開。

- 如雞只有兩斤左右，熄火後浸二十分鐘就可以。

- 下冰糖的作用是令雞肉更嫩滑。

- 勿太早放入玫瑰露進滷水料內，會揮發掉玫瑰露的香氣。

- 可用筷子在雞翼和雞髀之間輕輕戳一下，如能輕易戳穿代表已熟；如尚未熟，可放進滷水汁內再浸。

- 雞攤涼後才斬件，可避免在斬件時雞皮收縮或「甩皮甩骨」。

- 滷水汁可儲存在雪櫃，留待日後滷雞髀、雞翼、豬胜等等。

仙翁西米露

材料

仙翁米	1/4 兩
西米	半量杯
冰糖	4 兩
薑片	2-3 片

做法

1. 仙翁米用清水浸 1 小時至發脹，放入有薑片的滾水內氽水，瀝去水分，過冷河，瀝乾水分（約有 1 飯碗份量）。

2. 西米放入滾水內煮 10 分鐘，再焗 10 分鐘，過冷河，瀝乾水分。

3. 煲滾適量水，下冰糖，待糖溶後（須試味），倒入仙翁米，滾一會後下西米，待再煮滾即可享用。

鼎爺話你知

· 仙翁米是生長於淡水的野生藻類，水質一定要清澈無污染。經浸發後的仙翁米外型渾圓，呈墨綠色，仿如一顆顆絲珍珠。

· 在上世紀六、七十年代，仙翁西米露風行一時，壽星公的六十大壽筵席，常以仙翁米為甜品的材料；壽星婆的壽宴則享用皇母蟠桃包，各有不同的祝壽寓意。

· 因仙翁米是藻類，帶點腥味，所以要加薑片氽水去腥。

· 仙翁米浸發後要即日使用，如放在雪櫃儲存會吸收水氣，煮時會容易瀉身。

· 西米不用浸洗，放入滾水內攪勻，煮片刻再焗一會至透明，過冷河，這方法可令西米呈完整的顆粒。

· 這糖水可以加入椰奶或杏汁，同樣美味。

●未浸發的仙翁米

●已浸發的仙翁米

仙翁西米露

花膠陳皮紅棗
南北杏糖水

材 料

南杏	2 兩	┐
北杏	4 錢	├ 南北杏比例是 5：1
40 頭花膠	3 個	
陳皮	1 塊	
紅棗	8 粒	
冰糖	5 兩	
連皮薑片	3 片	
蔥	2 條	

做 法

1. 花膠用水浸 2 小時。凍水放入薑（連皮）和蔥煮滾，放入花膠，熄火，
   焗 2-3 小時至花膠軟身；撈起，切件。圖 1~3

2. 陳皮浸軟，刮去瓤。

3. 南杏、北杏沖淨；紅棗洗淨，去核。

4. 煲滾適量水，放入南北杏煲約 45 分鐘至腍，加入冰糖，待冰糖溶化後
   下陳皮和紅棗，煲 20-30 分鐘，試味。轉為大火，放入花膠，再滾後熄
   火，蓋上鍋蓋焗 5 分鐘，即可享用。圖 4~6

花膠陳皮紅棗南北杏糖水

鼎爺話你知

· 花膠要最後才放入，因為花膠煲太久會溶掉。

· 南北杏的比例是五比一，因北杏含微量毒素，故不宜多放。

文思芝麻糊

材　料

黑芝麻	半斤
白芝麻	少許
冰糖	適量
花生	1 碗（淡脆口花生）
白米	1 碗
馬蹄粉水	適量
滑豆腐	1 盒

做　法

1. 用白鑊炒香黑白芝麻，攤涼。圖 1~2
2. 用白鑊炒香白米。
3. 黑白芝麻、花生和炒米放入攪拌機，注入水，攪成糊狀，用煲湯魚袋隔去渣。
4. 煲滾芝麻漿，加入冰糖，待冰糖溶化後注入馬蹄粉水，攪勻成芝麻糊，放入文思豆腐，煞是美麗。

文思豆腐做法

文思豆腐是揚州名菜，將一磚豆腐切成幼如頭髮的豆腐絲，如花朵在碗裏綻放，味道清淡，外形素雅。除了要講求精湛的刀工外，同時要手定、眼定、心定，才能一氣呵成不把豆腐切斷。

鼎爺話你知

· 黑芝麻、炒米、炸花生與水的比例是一比一點五。

· 如有對花生敏感者忌食這甜品。

栗子腰果蟲草花素湯

腰果　　　　半斤

湘蓮　　　　3兩

花生　　　　4兩

新鮮栗子　　1 1/2 斤（去殼）

蟲草花　　　少許

杞子　　　　半兩

做 法

1. 腰果、湘蓮、花生分別浸約2小時，瀝去水分。蟲草花、杞子沖淨。

2. 把栗子放入滾水內，待再煲滾後，栗子衣會脫落，撈起栗子，備用。

3. 將全部材料（杞子除外）放入鍋內，注入適量水，大火煲滾，改小火煮至花生腍後才下杞子，需時約1 1/2小時，再煲約半小時即可飲用。

4. 先試味，飲湯時，才各自在湯碗下鹽調味。

鼎爺話你知

· 這湯有清腸胃、補腎的功效。

· 宜挑選輕身、乾爽的蟲草花。

三菌青邊鮑燉烏雞湯

材 料

青邊鮑	1 隻
姬松茸	半兩
羊肚菌	半兩
蟲草花	半兩
烏雞	1 隻
金華火腿	1 兩
雞腳	6 隻
瘦肉	半斤
粗鹽	適量

做 法

1. 烏雞洗淨，切大件；瘦肉洗淨，切大件；火腿切大粒。

2. 鍋內注入凍水，放入雞腳、烏雞、瘦肉、火腿汆水，洗淨。斬去烏雞和雞腳腳趾甲，再在肉枕剁一刀，令燉湯時骨膠原盡出。圖 1~3

3. 羊肚菌、蟲草花、姬松茸浸軟後洗淨；羊肚菌、姬松茸切蒂，備用。

4. 把鮑魚嘴切去，用粗鹽清洗鮑魚，洗淨，斜切鮑魚片（燉湯時容易出味）。

5. 燉盅內放入瘦肉、雞腳、部份鮑片、烏雞、羊肚菌、蟲草花、姬松茸、金華火腿和餘下鮑片，注入滾水，封上耐熱保鮮紙或紗紙燉 3 小時即成。圖 4~5

三菌青邊鮑燉烏雞湯

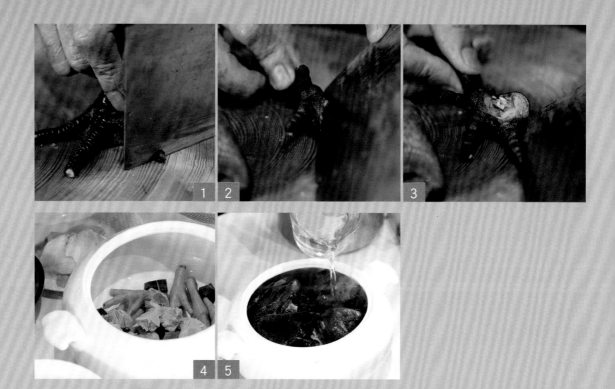

· 在南澳洲生長的青邊鮑大隻多肉，味道鮮甜，採用一隻已足夠燉湯。若用本地鮮鮑魚，味道略遜，但都適宜煲湯，可以改用本地生鮑魚兩隻。

· 羊肚菌宜去蒂部，因有苦味。

· 用作燉湯的鮑魚宜斜切片，燉湯時容易出味圖6；用作白焯的鮑魚宜片薄片。圖7~8

湯。茶

124

拆魚羹

材 料

大頭魚	1 條（約一斤）
木耳絲	1 大朵份量
甘筍絲	1 條份量
新鮮竹筍絲	2 個份量
勝瓜絲	半條份量
海參絲	1 條份量
薑絲	適量
唐生菜	2 棵
馬蹄粉水	
鹽	

湯 底

白鯽魚	1 條
鯪魚骨	6 條
柴魚	2 條
薑片	

做 法

1. 湯底做法：柴魚拆肉；白鯽魚、鯪魚骨加薑片煎香，注入水，下柴魚，加熱，熬成奶白色魚湯。撈起湯渣，挑出柴魚，細心撿走柴魚骨，柴魚肉備用。

2. 大頭魚起魚腩圖 1~4；魚頭破開、斬件，放在蒸碟上，鋪上薑絲，蒸 5 分鐘至魚頭六成熟時，放入魚腩，鋪上薑絲，續蒸 8 分鐘。

3. 魚頭、魚腩放涼，拆肉備用。

4. 將唐生菜的菜柄幼絲。圖 5~7

5. 湯底加熱，倒入魚肉、海參絲、木耳絲、甘筍絲、竹筍絲、勝瓜絲，下少許鹽調味，再滾起後，下馬蹄粉水打芡，加入唐生菜絲，趁熱享用。圖 8~9

湯
。
茶

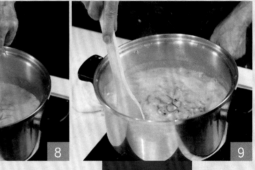

鼎爺話你知

· 這是個順德菜，順德
是漁米之鄉，以食河
鮮聞名。雖然河鮮鮮
味不及海鮮，但如懂
得烹調，一樣美味。

· 將柴魚先拆肉才熬湯，
會容易出味。柴魚是
湯底的靈魂，提升了
湯底的鮮味。

· 將生菜菜柄切成絲，
吃時非常爽口。

白果腐竹燉雞湯

材　料

雞	1 隻
瘦肉	半斤
金華火腿	適量
白果	3 兩
腐竹	1 張
炸枝竹	3 枝

做　法

1. 瘦肉切大件；雞斬大件，切去腳趾甲，與瘦肉、金華火腿一起汆水，沖淨。

2. 白果去殼，放入滾水內，看見白果開始脫皮，撈起過冷河，揉去白果衣。

3. 腐竹、炸枝竹浸軟；腐竹切成大塊，炸枝竹切成 3 吋長段。

4. 燉盅內放入炸枝竹、雞、瘦肉，用腐竹圍邊，放上白果和火腿，注入滾水至九成滿。蓋上燉盅蓋，包上耐熱保鮮紙或紗紙，燉約 3 1/2 小時即成。

鼎爺話你知

· 讀者可能會問：為甚麼不撕去雞皮？因為雞連皮煲湯，湯才香濃，如怕肥膩，應盡量切去雞的肥膏。

· 如只用雞、腐竹、白果煲湯，湯味會較淡，所以下瘦肉和金華火腿提味。

白果腐竹燉雞湯

清熱解毒茶

●金錢草

●雞骨草

●溪黃草

材料

雞骨草	4 兩
溪黃草	4 兩
金錢草	4 兩
蜜棗	8 粒

做法

1. 把雞骨草、溪黃草和金錢草用水沖一沖，再用凍水浸一夜。

2. 翌日將水倒去。注入清水，放入雞骨草、溪黃草和金錢草，汆水。

3. 撈起雞骨草、溪黃草和金錢草，放入鍋中，加入蜜棗和適量水，水滾後煲 3 至 4 小時即可飲用。

鼎爺話你知

· 雞骨草、溪黃草和金錢草所用的全是乾貨，在生草藥舖有售。

· 將雞骨草、溪黃草和金錢草浸一夜，除了浸去附着枝葉的泥沙外，還可以讓草藥快些出味。

· 這茶有清熱解毒之效，一年四季皆可飲用。

· 這材料份量適合六至八人飲用。

清熱解毒茶

芋頭豬油渣炒飯

材料

芋頭	3 兩
肥豬肉	半斤
白飯	2 碗
蝦米	1 兩
紅葱頭	適量

調味料

糖	適量
魚露	適量

做法

1. 蝦米用水浸軟，略剁備用。
2. 把芋頭切成方粒備用。
3. 肥豬肉切粒備用。
4. 起油鑊放入肥豬肉粒，待肥豬肉粒炸至油開始出時，可加入少量水以逼出豬油。待豬油渣炸至香脆後，撈起，瀝乾油備用。
5. 用豬油把芋頭炸至熟和香脆後，撈起，瀝乾油備用。附圖
6. 用豬油起鑊，爐火轉細些，加入部分紅葱頭炒香，倒入白飯推勻，下蝦米，炒勻後下炸芋頭粒，下少許糖調味，倒入餘下的紅葱頭，下魚露，兜勻後下豬油渣，趁熱享用。

鼎爺話你知

· 宜購買豬背的脂肪，炸後特別香脆。

· 將芋頭拉油除了較香口外，還可令芋頭在兜炒時不易散開。

芋頭豬油渣炒飯

豉椒蝦乾魚乾伴白粥

材料

蝦乾	2 兩
銀魚乾	1 兩
蒜頭	半個（剁蓉）
紅葱頭	4 粒（剁蓉）
豆豉	約 2 湯匙（略剁）
紅辣椒	2 條（切粒）
糖	適量
鹽	適量
老抽	適量（調色用）

粥料

新米	適量
腐竹	1 塊

●銀魚乾

●蝦乾

做法

1. 白米洗淨，瀝乾，放入滾水內煮至米粒綿滑，放入腐皮煲至溶化

2. 用油炸銀魚乾至鬆脆，盛起備用。

3. 蝦乾浸軟，撈起瀝乾水分。蝦乾走油。

4. 燒熱鑊，下油，順序下蝦乾、蒜蓉、紅葱頭蓉、銀魚乾和豆豉炒勻，倒入紅椒粒，下糖調味，炒勻後下鹽，兜炒至味道均勻，下少許老抽調色即可。

豉椒蝦乾魚乾伴白粥

鼎爺話你知

- 天時暑熱，食慾不振，宜用豉椒蝦乾魚乾伴白粥享用。

- 炸銀魚乾不用洗，揀去雜質便可以；因浸洗後的銀魚乾會變軟，除非將它再風乾。

- 炙在炸銀魚乾的途中會下少許水，會令銀魚乾炸得更鬆脆，但有一定危險，因爲熱油會四濺，要小心圖1～3。如水分太多，就會如圖般「爆炸」。圖4

- 我喜歡用新米煮粥，因爲新米水分多、米身軟，容易煲綿。

- 食譜內的材料份量可隨你喜歡調整。

炸油糍

材　料

白蘿蔔	半斤
紅蘿蔔	1 條
蝦米	1 兩
中筋麵粉	3 兩
粘米粉	半兩
發粉	半茶匙
葱粒	2 棵份量

調　味　料

胡椒粉	少許
五香粉	半茶匙
鹽	半茶匙
麻油	1 茶匙

做　法

1. 白蘿蔔、紅蘿蔔刨絲，白蘿蔔用手擠乾水分。

2. 蝦米剁幼粒。麵粉、粘米粉、發粉篩勻

3. 將白蘿蔔絲、紅蘿蔔絲和粉類拌勻，加入蝦米、五香粉、胡椒粉拌勻。注入適量水，調勻後下麻油，拌勻後下葱粒。圖 1~4

4. 熱鑊下油，將油糍模型放入鑊內浸一浸油，倒走多餘油分後，放入蘿蔔漿至八成滿，放回油中炸至離模，顏色金黃，撈起，瀝去油分。圖 5~7

· 蘿蔔絲不能切得
太幼，切這個粗
幼度就可以，否
則會出水。

· 每次舀入粉漿時
必須將模型浸入
油內燒熱，炸油
糍才容易鬆脫。

炸油糍

139

第一次目擊文思豆腐的誕生

在拍攝文思豆腐時,我們一眾的工作人員只見到鼎爺在不斷切豆腐,心想怎樣將豆腐變成綻放的花朵呢?當鼎爺叫我們:「睇住喇」,他用小木鏟將豆腐舀起,放在水裏淘一淘,再放在水中,奇妙的事情發生了,真的綻放成一朵花!

我們看到目瞪口呆,情不自禁地說:好犀利!

還有更犀利的,是舀起豆腐的小木鏟是鼎爺自製的。

除了文思豆腐,在拍攝期間鼎爺為我們示範怎樣切薄如蟬翼的蘿蔔片、冬瓜、蘿蔔網等等。

刀工的準繩與細緻,真是開了眼界。

PAN-FRIED JAPANESE EEL

Refer to p.11 for steps

INGREDIENTS
1.2 kg or more
Japanese eel
coarse salt
caltrop starch
spring onion
shallot
garlic
sliced ginger

MARINADE
ground pepper
chopped shallot
chopped garlic
grated ginger
rice wine

SEASONING
sugar
salt
rice wine

METHOD

1. Section spring onion. Pat garlic and shallot flat with a knife.

2. Rub Japanese eel with coarse salt, rinse with running water. Meanwhile, wipe off the slime with a scrub sponge. Rinse well. Soak up the water with a dry towel.

3. Cut the Japanese eel into thick pieces. Marinate for a while. Coat with caltrop starch. *Picture 1-3*

4. Heat up a wok and pour in oil. Put in the Japanese eel, pat away the excessive caltrop starch before pan frying. Pan fry on high heat until 70 % done. Place it on a plate and leave aside for later use. *Picture 4*

5. Heat up a casserole and pour in oil. Stir fry sliced ginger, garlic and shallot until fragrant. Put in half portion of the spring onion. Stir fry until soft. Add in the Japanese eel. Cover with a lid and cook for 1-2 minutes. Add sugar. Cover again for 30 seconds. Add spring onion and a pinch of salt. Cover again and leave for another 30 seconds. Turn off the heat. Pour rice wine on the lid. Serve. *Picture 5-7*

GRANDPA'S TIPS

- *Suggest using a Japanese eel that at least weight 1.2 kg so that you can enjoy a crunchy, meaty and firm mouthfeel.*
- *The best way to remove the slime from the eel is rub with coarse salt and scrub sponge. Wash it with hot water may be a risk as the meat will be getting old if you lost on time control. Make sure slime is removed completely, or the texture of the eel will be very soft and sticky like glue.*
- *The reason to pat flat garlic and shallot before stir frying is to allow them release more flavor that the dish will be more aromatic.*
- *Japanese eel is a freshwater fish which has fishy and muddy smells. Use marinade to remove the unpleasant smell and enrich its flavor.*
- *The secret to make a crispy Japanese eel with aromatic caramelized flavor is melt sugar before adding eel, in this way, the melted sugar will stick on the eel so that the eel will be caramelized lightly after heating.*

INGREDIENTS

1.2 kg beef interim brisket
1 white radish
3 tbsp Chu Hou Paste
10 or more ginger slices
113 g rock sugar

HERBS

6-7 star anise
4 pc Cao Guo
2 pc cinnamon
8 pc bay leaf
6 pc licorice
19 g white peppercorns
2 dried tangerine peels

METHOD

1. Peel white radish and roll cut. Slightly pat white peppercorns until it's broken.

2. Cut beef interim brisket into pieces. Pour in cold water to a wok. Put in the beef interim brisket, a few slices of ginger, approx. 9 g of rock sugar, 1/2 dried tangerine peel and a little white peppercorns. Don't cover the wok and just scald the beef interim brisket until there're blood and bubbles come out. Take it out from the wok. Rinse well and drain.

3. Heat up a wok and add oil. Put in the rest of the sliced ginger. Stir fry briefly and take it out.

4. Pour water into a casserole. Put in the stir fried sliced ginger from step 3, herbs and rock sugar. Cook until the rock sugar is dissolved. Put in the beef interim brisket. Stew for about 1 hour over medium heat. Test the beef brisket and see if it's tender enough. If so, put in Chu Hou Paste. Turn to low medium heat and stew for a while.

5. Put the white radish into the casserole. Stew for 20 minutes until it's tender. Serve.

BRAISED BEEF INTERIM BRISKET IN CHU HOU SAUCE

GRANDPA'S TIPS

- *Interim brisket is the meat which is near the ribs. After debone it, there will be a pit on the meat and that's why it's called Niu Keng Nan in Chinese. The word "keng" in Chinese means pit. Interim brisket has rich beef flavor and with no unpleasant smell. It's an ingredient that can be cooked for long time, therefore, it's a perfect ingredient for stew or soup.*

- *The taste of the Chu Hou Sauce is quite strong, and the aroma of the beef brisket would not come out in a short time. Therefore, don't add in the sauce too early, or it will hijack the major role from the beef brisket by covering its flavor.*

- *Use daikon (Japanese radish) if you can. As it is rich in moisture that the meat is soft and crisp without any root or membrane.*

- *Stir fry ginger briefly in hot oil to reduce the excessive moisture from the ginger in order to help the ginger soak in the aroma of the beef and the sauce. Thus, the ginger will be very tasty.*

POACHED PIGEON WITH DRIED TANGERINE PEEL

Refer to p.17 for steps

INGREDIENTS

2 pigeons
75 g dried tangerine peel
demerara sugar

METHOD

1. Gut and rinse pigeon. Pour cold water into a wok. Put in the pigeon and don't cover the wok. Blanch the pigeon until its blood is released. Rinse and drain well. Leave aside for later use. *Picture 1*

2. Soak dried tangerine peel until soft. Scrape off the pith. Cut into large pieces. *Picture 2*

3. Pour water (the quantity should be enough to cover the pigeon) into a casserole. Add in the tangerine peel and demerara sugar. Cook for 1 hour. Put in the pigeon. Cook for 3 minutes over high heat. Turn off the heat. Leave the lid on the casserole and cover for 40 minutes. Serve. *Picture 3-4*

GRANDPA'S TIPS

- *What is dried tangerine peel? Dried tangerine peel is the peel of Xinhui mandarin that has been sun-dried and kept for at least 5 years. The one which hasn't been kept for 5 years is called citrus peel or tangerine peel.*
- *Why I use casserole instead of stainless steel pot? It's because casserole can maintain heat that the pot of food could remain in a proper temperature after set still for 40 minutes.*

BRAISED AMUR CATFISH IN SHUNTAK STYLE

Refer to p.20-21 for steps

Refer to p.20-21 for steps

GRANDPA'S TIPS

- *Try to buy Amur catfish which has small head and round short body.*
- *Only put in shrimp after the calm shell is opened, or the shrimp will be over-cooked. Also, don't cover the casserole after putting in shrimp, otherwise the texture of shrimp will become rough.*

INGREDIENTS

1 Amur catfish
300 g roasted pork belly
300 g clams
150 g shrimps
caltrop starch
sliced ginger
shallot
garlic
sectioned spring onion
1/4 green bell pepper
1/4 red bell pepper
1/4 yellow bell pepper

SEASONING

1 tbsp sugar
1 tsp salt
1 tbsp light soy sauce
dark soy sauce

METHOD

1. Cut bell peppers into wedges. Slice roasted pork belly. Rinse clams after they release their sands.

2. Rinse Amur catfish and soak up the water. Make a few marks on one side with a knife slightly. Thinly coat with caltrop starch. *Picture 1-3*

3. Heat up a wok and pour in oil. Pat away the excessive caltrop starch from the fish. Shape it into crescent-shaped. Deep fry until half done. Take it out from the wok. Drain. Remember to put the fish head into the wok first when deep frying, as it's difficult to get cooked. Also, keep scooping the boiling oil to the back of the fish to make sure it get cooked easily. After taking out the fish, scoop the boiling oil on it once or twice. *Picture 4-6*

4. Heat up a casserole. Pour in oil. Stir fry sliced ginger, shallot and garlic until fragrant. Add in the sliced roasted pork belly. Cook until its oil comes out. Put in the clams. Stir fry well. Put a splash of water and sectioned spring onion into the casserole. Cover with a lid and cook until the calm shells are opened (take about half minute).

5. Put in shrimps and don't cover. Cook until the shrimps turn red in color. Add in the bell peppers. Stir well and season with sugar and salt. Take out a few portion of the ingredients. Add in the Amur catfish. Put the ingredients back. Add a few drops of dark soy sauce for coloring. Cover with a lid. Cook for 5 minutes on medium high heat. Turn off the heat. Season with light soy sauce. Cover and leave for 1 minute. Serve.

STEWED CATFISH WITH BLACK BEANS

*Refer to p.25 for steps

INGREDIENTS

2 catfishes
300 g black beans
1/4 dried tangerine peel
8 red dates
1 stalk garlic sprout (sectioned)
chopped shallot
diced ginger
caltrop starch
coarse salt

SEASONING

1 tbsp Chu Hou Paste

METHOD

1. Soak dried tangerine peel until soft. Rinse red dates and seed.

2. Toast black beans in a wok without oil until its skin is broken; and you can see the cyan color inside. Add water and boil until it's boiling. Drain the black beans with a colander.

3. Put a proper quantity of water; black beans; dried tangerine peel and red dates into a casserole. Boil for 40 minutes.

4. Meanwhile, use coarse salt to wipe off the slippery liquid from the fish. Rinse and use a dry towel to soak up the water. Pat some caltrop starch on the fish, then deep fry it until it's half cooked. *Picture 1*

5. After the black beans become soft, add the fish into the pot. *Picture 2* Stew for 10 minutes, now the fish supposedly soaked in the sauce completely. Add the garlic sprout, chopped shallot and diced ginger into the pot. Cover with a lid and stew for a while. Turn to medium heat and add Chu Hou Paste. Taste and stew for a while. Serve.

GRANDPA'S TIPS

- *Black bean that with good quality is called Hei Pi Qing. The color of its meat is cyan. Beans are always come with unpleasant smell, and to get rid of the smell, stir fry it without oil until its skin is breaking out before cook. As a result of doing this, the bean would be more crispy; sweet and flavorsome.*

- *There are two types of catfish in the market, one is go with grey color and the other one is sand yellow. Buy the sand yellow one, as the texture is more smooth and tender, the taste is more sweet with no muddy taste.*

- *The reason to pat caltrop starch on the fish and deep fry it before cooking, is to make a firmer texture so that its shape can be retained. Also, it will be more fragrant and sweet.*

- *Deep fry the fish to half cooked instead of cooked so that it can soak in the sauce while stewing.*

- *When Grandpa Steve is a little child, his fellow townsfolk used to make this dish for health enhancement as this dish can enrich the blood.*

STEWED SMALL SNAKEHEAD SOUP WITH HUAI SHAN AND GOU QI ZI

Refer to p.28 for steps

*Refer to p.28 for steps

GRANDPA'S TIPS

- *Small snakehead can nourish your organs; improve health; enrich blood and spleen; and strengthen the immune system.*

- *Scald small snakehead can remove the slime on the fish. To dissolve the fishy smell, scald it with sliced ginger.*

- *Soup made with small snakehead and ingredients such as dried Longan and ham, taste clean and sweet; and is not greasy. This soup is rich in nutrient that is good for women who gave birth by C-section as it assists in post-cesarean wound healing.*

INGREDIENTS
2 small snakeheads
300 g lean pork
HKD20 in total Huai Shan; Gou Qi Zi and dried Longan
Jinhua ham

METHOD
1. Soak Huai Shan in water for an hour. Rinse well. Rinse Gou Qi Zi and dried Longan.

2. Gut small snakeheads and rinse well. Scald with sliced ginger. *Picture 1-2*

3. Rinse lean pork and cut it into large dices. Scald with ham.

4. Put all ingredients into a large slow cooker. Pour boiling water into the cooker. Seal with microwave wrap or mulberry paper. Stew for 3 hours. Serve. *Picture 3-6*

FISH SOUP WITH PEANUTS, BLACK-EYED PEA AND PAPAYA

Refer to p.31 for steps

INGREDIENTS

1 big fish tail
300 g peanuts
150 g black-eyed peas
1 green papaya
8 chicken feet
300 g lean pork
4 slices ginger

METHOD

1. Soak peanuts and black-eyed peas in water overnight. Rinse well. *Picture 1*

2. Wash fish tail and wipe dry. Heat up a wok and pour oil into it. Stir fry sliced ginger until fragrant. Put the fish tail into the wok and pan fry until fragrant.

3. Peel and seed papaya. Scrape the piths off from it. Then cut it into large pieces. *Picture 2-4*

4. Cut lean pork into large pieces. Scald with chicken feet. Rinse well. Cut the nails off from the chicken feet and make a light mark on the sole of the feet with knife. *Picture 5-6*

5. Pour a proper quantity of water into a pot. Boil until it's boiling. Add peanuts and wait until the water is boiling again, then boil for 20 minutes. Add black-eyed peas, chicken feet and lean pork. Boil for another 20 minutes. Put papaya into it and boil for about 30 minutes. Add the fish tail. Turn to medium heat and boil for an hour. Serve. *Picture 7-8*

GRANDPA'S TIPS

- *The functions of this soup are to induce lactation; strengthen the spleen; benefit the stomach; nourish the skin, loosen the bowel and release constipation.*
- *The piths of papaya must be removed when preparing, or it will cause a bitter taste in the soup.*
- *The texture of papaya is soft and tender, therefore, don't put it in to the soup at the beginning, instead, add it in the middle of the cooking to prevent it melts completely.*
- *The reason to make a light mark on the chicken feet sole is to let collagen released from it when boiling the soup.*
- *Add salt to each bowl of soup while taking the soup instead of putting it into the pot, since the soup will taste sour after re-heating on the next day.*

BRAISED PORK TROTTER WITH GINGER AND SWEET VINEGAR

Refer to p.35-36 for steps

INGREDIENTS
3 kg sweet vinegar
150 g black rice vinegar
300 g Shuang Zheng rice wine
900 g mature ginger
1.8 kg ginger (half-mature)
1.2 kg trotter
16 eggs

TOOLS
1 bamboo mat (washed)
large clay pot

METHOD

1. Rinse and peel the ginger. Pat it with a knife until flat. Air-dry it outside, or toast it in a wok without oil until its water is evaporated. *Picture 1-4*

2. Ask the butcher to cut the trotter into large pieces. Rinse it after back home. Put it into a cold water, turn the heat on and boil until the water is boiling. Press until blood and dirt come out. Take it out from the pot. Rinse with cold water.

3. Bring a pot of water to a boil. Put the trotter into it and boil for 10 minutes. Take it out and rinse under cold water again. Drain well.

4. Put the bamboo mat in the clay pot. Pour sweet vinegar into it. Bring to a boil, then cook for another 20 minutes. Put in ginger and turn to low heat; and cook for 40 minutes. Pour black rice vinegar. Boil for 1 hour 30 minutes then put the trotter in. Pour Shuang Zheng rice wine. Boil over medium low heat for 2 hours. Turn off the heat, cover and leave it overnight. *Picture 5-7*

5. Re-heat the pot of trotter before eating. Put in the boiled and shelled eggs. *Picture 8* Boil for 10 minutes. Turn off the heat and leave it until all ingredients soak in the taste completely. Serve.

GRANDPA'S TIPS

- This dish is an indispensable health food for new mum who's at the postnatal period (the 1st month after giving birth). Woman who had natural childbirth can have this dish after 12 days of the delivery. Women who gave birth through C-section should wait until blood stasis clears up before taking this dish.

- To let the ginger soaks in the sweet vinegar flavor completely, air-dry or toast dry it before using so that the ginger would taste better than trotter and egg.

- Suggest you scrape off the epidermis of ginger instead of peeling its whole skin, as a result of doing this, the ginger will be more flavorful. *Picture 9-10*

- Mature ginger and half-mature ginger look alike that it's difficult to recongnize them. Half-mature ginger is a ginger that the planting time is in-between baby ginger and mature ginger. Pull open the ginger and look at it, if it comes with no root and looks smooth which means it is half-mature ginger. *Picture 11* As mature ginger always comes with lots of roots and it's because it had been planted for long time. *Picture 12*

- Scald trotter twice to enrich both the taste and the texture. The first time is to remove its blood and dirt. *Picture 13-14*. At the second time, you have to put the trotter into hot water and boil for 10 minutes, then rinse with cold water, the purpose is to get a crispy skin and make the meat tender. *Picture 15-17*

- You may wonder what is the purpose of using Shuang Zheng rice wine. It's because the wine can remove the unpleasant smell from the meat. Also, the wine won't affect the flavor of the trotter, as it would have evaporated during cooking.

- To prevent trotter sticks in the pot, put a bamboo mat in the clay pot before placing any ingredient.

CLAY POT RICE WITH FROGS

Refer to p.39 for steps

Refer to p.39 for steps

GRANDPA'S TIPS

- *Frog is good for woman whose Yuan Qi (constitution) was badly damaged after pregnancy and giving birth, as its functions is to strengthen the Yuan Qi and invigorate the spleen. If the new mum in your family is lack of energy; easy to get tired and loss of appetite, you may cook this clay pot rice for her.*

- *To prevent food poisoning, make sure the frog is well done before eating.*

- *Don't pile up the frogs when arrange them on the rice, or you will have trouble to get them cooked completely.*

- *In step (3), regarding the proportion of rice to water, it is 1 to 1 but I suggest reducing the quantity of water a bit as the frog will release water while cooking. Therefore, to prevent the rice becomes too soft and wet, adjust the quantity of water accordingly.*

INGREDIENTS
4 frogs
rice
(1/3 new rice, 2/3 old rice)

MARINADE
caltrop starch
oil
sugar
chopped shallot
ginger juice
wine

SWEET SOY SAUCE (BOIL TOGETHER UNTIL SUGAR MELTS COMPLETELY)
light soy sauce
dark soy sauce
cooked oil
sugar

METHOD

1. Gut frogs, remove the back bones and cut off the feet. Cut into pieces. Make a few marks on the thick part of the legs with a knife slightly so that you can heat up the frog legs evenly. *Picture 1*

2. Marinate the frogs with caltrop starch, oil, sugar, chopped shallot, ginger juice and wine one by one in order. Leave aside for 20 minutes.

3. Wash and rinse rice, put it into a clay pot. Pour water, the proportion of rice to water is 1 to 1 (can reduce the quantity of water a bit). Boil over high heat until the water is boiling. Turn to low heat and boil until you see bubbles rising in the water. Put in the frogs and turn the heat up a bit. Cook until you hear a sound of "la la" ,and smell the fragrance of the rice which means the rice begins to get burnt. Turn to low heat again and boil for another 15 minutes. Turn off the heat. Sprinkle diced spring onion and pour sweet soy sauce on top. Toss the rice until it loosens. Cover and leave for 5 minutes. Serve. *Picture 2-4*

CLAY POT RICE WITH DRIED PERSIMMON

Refer to p.41 for steps

INGREDIENTS

3 dried persimmons
rice

METHOD

1. Cut each of the dried persimmon into 3 pieces. Put aside for later use.

2. Put rice and water into a clay pot, the proportion of rice to water is 1 to 1. Put half portion of the dried persimmon into the pot. Stir well. Cover with a lid and boil over high heat until the water starts to evaporating. Then put the rest of the dried persimmon into it. Turn to low heat. Cover again with a lid, pay attention to the sound of "la la" from the pot; and until the fragrance of the rice is released. Turn off the heat and cover for 5 minutes. Serve. *Picture 1-4*

GRANDPA'S TIPS

- *This dish can invigorate the spleen and improve appetite.*
- *Dried persimmon can be bought at Chinese medicine shop. The white powder on the surface is persimmon frost, please leave them on the persimmon as they are cold-natured and taste sweet. Also, they can dissolve Heat; moisten Dryness and reduce phlegm.*

JOB'S TEARS, DANZHUYE AND PORK SHOULDER BLADE SOUP

INGREDIENTS

38 g Job's tears
38 g cooked Job's tears
1 bundle Danzhuye
600 g pork shoulder blade
19 g Medulla Junci
1 corn on the cob

METHOD

1. Scald pork shoulder blade and rinse well.

2. Soak Job's tears in water for an hour and rinse well. Rinse Danzhuye and Medulla Junci. Rinse corn and cut into large pieces.

3. Pour a proper quantity of water into a pot. Put all ingredients into it. Turn on the heat and boil over high heat for 20 minutes. Turn to medium heat and cook for 1 1/2 hours. Serve.

GRANDPA'S TIPS

- *This soup has light bamboo leaves flavor which is different from other soups made from meat.*
- *In the past, people who lives in village will make this soup for young child who is Heatiness (Yeet Hay) and has yellow urine.*
- *Cut corn on the cob into large pieces so that its flavor can be released faster.*

STEAMED SLICED GROUPER WITH ANGLED LUFFA

*Refer to p.48 for steps

GRANDPA'S TIPS

- *The outer ridges of angled luffa is very hard, therefore, must peel them off before cooking. Since the old skin has bitter flavor, it should be scraped off before cooking.*

- *Before slice the grouper belly, touch it with your hands to make sure there's no more bone inside. Or you have to remove the bone with a fishbone plier.*

- *Add some caltrop starch into the fish when marinating to prevent the dish from being watery after steaming.*

INGREDIENTS

1 angled luffa
900 g grouper belly
approx. 2 tbsp chopped garlic
diced spring onion
diced red chili
boiling oil
light soy sauce

MARINADE

oil
sugar
salt
caltrop starch

METHOD

1. Rinse grouper belly and slice. Mix with a little oil. Add sugar; salt and caltrop starch. Mix well. Leave aside for later use. *Picture 1*

2. Peel off the outer ridges of angled luffa, rinse it under running water; meanwhile, use a small knife to scrape off the old skin. Thinly slice the angled luffa at angles. *Picture 2*

3. Arrange the sliced angled luffa on a plate. Spread the sliced fish on top of it. Sprinkle the chopped garlic on top, steam for about 8 minutes. *Picture 3-4*

4. Sprinkle diced spring onion and diced red chili on top. Pour boiling oil over the fish. Add light soy sauce from the side. Serve when it's hot. *Picture 5-6*

STUFFED BEAN CURD PUFF WITH SHRIMP PASTE

Refer to p.51 for steps

INGREDIENTS

12 bean curd puffs
300 g shrimps
38 g fatty pork meat (finely diced)
1 calamansi
sugar
salt
caltrop starch
water
oil

METHOD

1. Shell and gut shrimps. Rinse and soak up the water with a dry towel. Wrap and roll with a towel. Put into fridge and leave for an hour.

2. Pat the shrimps flat and dice it. Pat with the back of a knife until it becomes mushy. Stir in clockwise direction until it's sticky. Add sugar and stir well. Stir in salt, fatty pork, a splash of water, a little caltrop starch and a few drops of oil one by one; completely; in order.

3. Cut open one-third of the bean curd puff from the top. Turn the puff inside out. Stuff shrimp paste into it. *Picture 1-2*

4. Heat up oil, make sure the temperature is not too high. Put in the stuffed bean curd puff gradually, with the shrimp paste side down. Deep fry until slightly brown. Turn to high heat to release oil from the bean curd puff. Take the puff out from the wok. Drain. *Picture 3-6*

5. You may squeeze some calamansi juice on the stuffed bean curd puff to enhance the flavor when serve.

GRANDPA'S TIPS

- Use fresh frozen shrimp for this dish would be fine enough. Follow step 1 to handle shrimp can soak up the excessive moisture from the shrimp so that you can enjoy a crunchy mouthfeel.
- Using the back of a knife to pat shrimp can prevents the shrimp becomes too gluey so that you can enjoy a nice mouthfeel. The shrimp paste would have lost the texture if it's too mushy.
- Add some fatty pork meat into shrimp paste to make it more flavorsome so that the shrimp paste will come with a silky smooth mouthfeel. Also, the texture will become solid but not too firm.
- Don't add in water chestnuts as it has a lot of moisture that will make the texture of the shrimp paste becomes loose after cooked.
- Want to present the dish in an elegant way? Use a scissor to make a few cuts on the side of the puff after stuffing in shrimp paste so that it looks like a flower.
- Don't turn off the heat while taking out the bean curd puff from the wok, or it will absorb the oil through.

GRANDPA'S HOMEMADE XO SAUCE

Refer to p.54 for steps

INGREDIENTS

150 g dried scallops
150 g dried shrimps
19 g Jinhua ham (finely diced)
6 cloves shallot
2 cloves garlic
7 red chilies
sugar
dark soy sauce

METHOD

1. Soak dried scallops until soft. Tear away the hard membrane, then thinly shred it. Soak dried shrimps until soft, finely chop it.

2. Finely chop shallot and garlic. Cut red chili into dices.

3. Heat up a wok with oil. Stir fry the chopped garlic until fragrant. Put in the chopped shallot and stir fry well. Add diced Jinhua ham into it. Put in the chopped dried shrimp. Stir well. Add in the dried scallop shreds, loosen the scallop shreds with a spatula to prevent the shreds stick together. Pour in a splash of water. Stir fry on high heat until the water is evaporated and the flavors of the ingredients released. Season with a little sugar and taste (must). Add in dark soy sauce for coloring. Stir well and put in the red chili. Stir fry until its color changed. Serve. *Picture 1-7*

GRANDPA'S TIPS

- *Wear a pair of gloves when cutting red chili to prevent your hands get the burnt feeling.*
- *Taste the Jinhau ham after you bought it home so that you can adjust the flavor more accurately when use it for seasoning.*

STIR FRIED PRAWN WITH HOMEMADE XO SAUCE

*Refer to p.57 for steps

INGREDIENTS

12 prawns
homemade XO sauce
1 stalk Chinese scallion
1/4 red bell pepper
1/4 yellow bell pepper
1/4 green bell pepper
caltrop starch solution

MARINADE FOR PRAWN

sugar
oil
caltrop starch
salt

METHOD

1. Shell prawns and soak up the water from it. Cut open the prawns in the middle and gut them. Make a light cut on each side of the prawn. Marinate with sugar, oil and caltrop starch for a while. *Picture 1-4*

2. Cut bell peppers into pieces. Oblique cut Chinese scallion into 2cm thick slices. *Picture 5-6*

3. Heat up a wok and pour in oil. Put in homemade XO sauce, half portion of the scallion and a little sugar. Cook until the color of the scallion changed. Put in prawns and stir fry well. Add a pinch of salt and the rest of the scallion. Put in the bell peppers and caltrop starch solution. Stir well. Take the ingredients out from the wok. Serve when it's hot. *Picture 7-10*

GRANDPA'S TIPS

- *Why I divide the Chinese scallion into 2 portions and add into the wok separately? The first time is for getting its flavor; and the second time is for a crunchy and crispy mouthfeel.*
- *You can adjust the quantity of homemade XO sauce according to your own taste.*

STEAMED EGG
WITH ASIAN
CLAMS

Refer to p.59 for steps

INGREDIENTS

3 eggs
12 Asian clams
diced coriander
diced spring onion
cooked oil
light soy sauce

METHOD

1. Pick clams' meat out from their shells after they released the grits. *Picture 1-2*

2. Crack eggs into a big bowl. Add in the clam meat. Stir well. Pour in water (the proportion of the egg to water is 1 to 1.5). Stir well. Pour the liquid into a steaming plate. Scoop out the bubbles. Wrap the plate with microwave wrap.

3. Bring water in a boil on high heat. Steam the egg mixture for 10 minutes. Take it out from the wok. Remove the wrap and sprinkle some finely diced coriander and spring onion on top. Pour cooked oil and light soy sauce on top. Serve when it's hot.

GRANDPA'S TIPS

- *Apart from salt, the good way to push clams release their sands is pour boiling water with about 60°C in temperature on it. Thus, the clams will release sands faster.*

- *After pouring the egg mixture into a steaming plate, stir the mixture with chopsticks to prevent the calm meat piling up.*

- *To make sure the steamed egg is silky smooth, remove the bubbles from the egg mixture is necessary before steaming.*

STIR FRIED RAZOR CLAM
WITH BROAD BEAN PASTE

Refer to p.63 for steps

INGREDIENTS

600 g razor clams
1 tbsp broad bean paste
1/4 red bell pepper
1/4 yellow bell pepper
1/4 green bell pepper
1 stalk Chinese scallion
broad beans

SEASONING

sugar
salt

THICK SAUCE

caltrop starch solution

METHOD

1. Thinly slice Chinese scallion at angle. Cut bell pepper into wedges. Finely press and grate broad beans, chop briefly and leave aside for later use. *Picture 1-3*

2. Cut razor clams from the middle with a knife. Gut and rinse well. *Picture 4-6*

3. Bring water to a boil. Put in the razor clams and scald. Take out from the pot when the meat is reducing in size and falling apart from the shells. Put into a bowl of ice water to cool down. Take out from the water. Soak up the water with a dry towel. *Picture 7-8*

4. Heat up a wok with oil. Put in half portion of the Chinese scallion. Stir fry until its color changed and turned to transparent. Add in broad bean paste, broad beans and the bell peppers. Stir well and put in the rest of the Chinese scallion. Add razor clams. Stir fry well. Add sugar, then salt. Pour in caltrop starch solution. Stir well. Serve.

GRANDPA'S TIPS

- Quantity of the fresh broad bean is depends on your own taste. Add some broad beans into the dish is to enrich the aroma and texture. It's because the broad beans in the broad bean paste had been ground completely that you can't find a whole broad bean.

- A lot of people only take the white part of the Chinese scallion. However, the green part is rich in scallion's flavor, therefore, don't discard it. Thinly slice Chinese scallion at angles is to help keeping the shape when stir frying in order to retain the aroma.

- The pith of bell pepper must be removed, or it would release a lot of water.

INGREDIENTS

6 dace fishes
10 stalks Chinese long beans
2 dried black mushrooms (soaked and finely diced)
38 g dried shrimps (soaked and finely diced)
38 g lean and fat pork (finely diced)

INGREDIENTS FOR BEANS BLANCHING

sugar
cooked oil

SEASONING

salt
caltrop starch
sugar
cooked oil

CALTROP STARCH MIXTURE

caltrop starch solution
oyster sauce

STUFFED CHINESE LONG BEAN WITH HAND CUT MINCED DACE FISH

Refer to p.66-67 for steps

METHOD

1. Cut the meat off from the fish ridge. Thinly slice 1/3 portion of the meat and chop briefly. For another 1/3 portion, scrape the meat off from the skin. Slice the remaining portion with skin on. Finely chop. *Picture 1-3*

2. Put fish meat, pork, dried black mushrooms and dried shrimps into a big bowl. Mix well. Add in salt, caltrop starch, sugar and mix well. Add in a splash of cooked oil, if the mixed meat paste is too dry you can add a few water into it. Toss the paste in the bowl for several times until it's sticky and firm. *Picture 4-5*

3. Remove the top of the Chinese long beans. Boil in boiling water until soft. When its color changed, add in sugar. Cook until it begins to soften. Pour in cooked oil. Cook until the beans is tender enough. Take it out from the pot and leave to cool. *Picture 6-7*

4. Make the Chinese long beans into ring-shaped. Stuff in the dace paste. *Picture 8-15*

5. Pour a proper quantity of oil into a pan. Pan fry the stuffed Chinese long beans until fragrant. Place on a plate. *Picture 16*

6. Boil oyster sauce with caltrop starch solution together into thick sauce. Pour on top of the stuffed Chinese long beans. Serve when it's hot.

GRANDPA'S TIPS

- *Use 3 different ways to cut the dace fish so that can enjoy 3 different mouthfeels in one single dish. Want to know how does it feel? Let's try it yourself.*

- *You may use knife or stainless steel spoon to scrape off the fish meat. To prevent the fish meat comes with bones, you must scrape the meat from the tail instead of head so that the bones will retained on the skin, or your throat would be hurt by the bones while eating.*

- *Chinese long bean should be thin and in even size so that you can make a beautiful ring.*

- *Add in sugar while boiling the long bean is to remove the grass smell and keep the green color. Also, put in oil can keep the color bright and fresh.*

DEEP FRIED GRASS CARP IN CHRYSANTHEMUM SHAPED

Refer to p.71 for steps

INGREDIENTS
1 big grass carp ridge
caltrop starch

MARINADE
salt
sugar

SAUCE
1/2 cup orange juice
2 tbsp lemon juice
1 tbsp demerara sugar
green peas
grated carrot

DECORATION
pine nuts (baked)
diced orange
finely diced red bell pepper
finely diced yellow bell pepper
finely diced green bell pepper

METHOD

1. Rinse grass carp ridge. Soak up with a dry towel. Make some stripes on the fish with a knife. Turn the fish in 45 degree and make another couple of stripes to form checker patterns. Oblique cut (towards inside) the fish meat into square slices. Marinate with salt and sugar for a while. Coat a proper quantity of caltrop starch. Leave aside for later use. *Picture 1-7*

2. Heat up a wok and pour in oil. Pat away the excessive caltrop starch from the fish. Put the fish sliced into the wok one by one. Stir fry until cooked and becomes chrysanthemum shaped. Take out from the wok. Drain and set it on a plate. *Picture 8-10*

3. Pour in a splash of water into another wok. Put in green peas and cook for a while. Add in grated carrot and pour in orange juice and lemon juice. Season with demerara sugar. Add a little caltrop starch solution into it and cook until the sauce thickens. *Picture 11-13*

4. Pour the sauce on the fish. Set the plate with the decoration. A colorful and mouth-watering dish is ready to be served. *Picture 14-15*

GRANDPA'S TIPS

- *Grass carp can be replaced by giant grouper or huge mandarin fish, however, their price are a bit expensive.*

- *The reason to oblique cut the fish into square slices is to make its shape into inverted trapezoid, in this way, the fish meat would roll up likes a chrysanthemum when deep frying.*

STEAMED MARBLE GOBY WITH PORK AND SOUR PLUM

Refer to p.75 for steps

INGREDIENTS

1 marble goby (600 g or more)
3 sour plums
2 dried black mushrooms
38 g tenderloin
pickled mustard green
sugar and cooked oil

METHOD

1. Soak dried black mushrooms until soft. Shred. Shred tenderloin and blanch. Leave aside for later use.

2. Seed and finely chop sour plums. Mix well with shredded dried mushrooms, shredded tenderloin and pickled mustard green. Add in sugar and stir well. Pour in a little cooked oil. Stir well and add a pinch of salt.

3. Gut and rinse marble goby. Drain. Cut open and remove the bone. *Picture 1*

4. Put a pair of chopsticks on the steaming plate, one for each side. Arrange the marble goby on it, pull open the fish stomach to make it lie on the chopsticks with the skin up. Add in the mixed shreds with chopped sour plum. Steam over high heat for 9 minutes. Pour in cooked oil. Serve. *Picture 2*

GRANDPA'S TIPS

- *Marble goby is firm and smooth in texture, sweet in taste and with no muddy flavor that is fit for steam or poach in oil.*

- *To make sure tenderloin is cooked, blanch it before steaming with fish as the steaming time for the fish is short.*

- *Debone the fish and lie it on a pair of chopsticks so that you can heat the fish evenly.*

STIR FRIED CRAB MEAT WITH DRIED SCALLOP AND VEGETABLES

Refer to p.78-79 for steps

INGREDIENTS

4 egg yolks
10 dried scallops
1 red crab
2 fresh bamboo shoots
38 g bean sprouts
38 g sugar snap peas
3 dried black mushrooms
salt
sugar

METHOD

1. Soak dried scallops until soft and shred it. Coat half portion of the dried shredded scallops with caltrop starch. Deep fry until crispy and fragrant.

2. Soak dried black mushrooms until soft and shred. Tear off the hard membrane from sugar snap peas. Shred the peas. Shred bamboo shoot and scald.

3. Steam red crab until cooked. Pick the meat off from the shell.

4. Blanch bean sprouts. Drain.

5. Stir egg yolks well.

6. Heat up a wok and pour in oil. Put in the dried scallops. Toss the shredded dried scallops to prevent it sticks together. Add in mushrooms, bamboo shoot and sugar snap peas. Add sugar and salt. Pour in the egg yolk liquid. Push and stir the liquid gradually until some dices formed like Osmanthus-shaped. Add in the bean sprouts and the deep fried dried scallop from step 1. Stir well. Serve when it's hot. *Picture 1-3*

GRANDPA'S TIPS

- *This dish is such a good cooking test for you. It tests your stir frying skill, the egg will get burnt if you turn the heat too high. However, if you turn the heat too low, the texture will become mushy.*

- *I suggest you use eggs that produced in Hu Bei province of China.*

- *Since bean sprouts will release water, blanch it before stir frying but don't blanch it too long, or it will become very tender and the crunchy texture will be lost.*

- *There's a secret for picking off crab meat, let me show you how to do it. See P.79*

- *You have to blanch the fresh bamboo shoot after shredding it to remove the bitter taste. Let's see how I handle it. See P.79*

STEWED PORK FINGER RIBS WITH PUMPKIN IN BLACK BEAN AND GARLIC SAUCE

Refer to p.82 for steps

GRANDPA'S TIPS

- *Pumpkin with heavy weight is better. I love using Chinese pumpkin, because it's more dense in texture and with less moisture that it won't become watery after cooking. Thus, you can enjoy a delicate and smooth mouthfeel.*

- *Don't over chop fermented black bean or pumpkin will be covered by black color after stewing. Also, the presentation is not good.*

- *Pork finger ribs is the meat in-between the rib and the meat that is soft, smooth and rich in pork flavor. I suggest you use Iberico pork ribs as the texture of it is more tender and smooth.*

- *The reason to cook pumpkin with skin on is because of its nutrition and nice flavor, though it's hard.*

INGREDIENTS
450 g pork finger ribs
900 g pumpkin
approx. 1 tbsp fermented black beans
garlic
shallot

MARINADE
sugar
oil
salt
caltrop starch

METHOD

1. Rinse and cut pumpkin into pieces, with skin on. Steam for 25 minutes. Leave aside for later use.

2. Rinse and drain well pork finger ribs. Marinate with sugar, oil, salt and caltrop starch one by one in order.

3. Chop fermented black beans. Pat garlic and shallot flat. Slightly cut.

4. Heat up a wok and pour in oil. Put in pork finger ribs and stir fry. Add a little sugar and stir fry for a while. Add in shallot and garlic. Stir well. Put fermented black beans into it. Stir well. Pour in water and season with a splash of dark soy sauce. Cover with a lid and boil for 15 minutes. *Picture 1-3*

5. Put in pumpkin and a little salt. Stir fry over low heat until the moisture of the pumpkin and the pork finger ribs are evaporated. Serve. *Picture 4-5*

GRANDPA'S PAN FRIED LAMB RACK

Refer to p.85 for steps

INGREDIENTS

1 lamb rack
2 tbsp cumin powder
1 tsp five spice powder
1/2 tsp salt
1 tbsp oil
1 clove garlic
3 shallots
2 red chilies
1/4 yellow bell pepper

METHOD

1. Soak up the water from lamb rack. Cut the lamb rack into pieces. Mix it well with cumin powder and five spice powder. Add salt and mix well. Pour in oil. Marinate for 10 minutes. *Picture 1-2*

2. Cut garlic, shallot, red chili and yellow bell pepper into dices.

3. Heat up a wok and pour oil into it. Put the garlic into the wok and deep fry until it turns light brown. Then, add diced shallot; diced red chili and diced yellow bell pepper one by one in order. Take them out from the wok and leave aside for later use. *Picture 3-4*

4. Heat up a pan and pour oil into it. Pan fry the lamb rack until fragrant. Put it on a plate. *Picture 5-6*

5. Arrange the deep fried dices from step 3 on top of the lamb rack. Serve when it's hot.

GRANDPA'S TIPS

- *Suggest using lamb rack which imported from New Zealand as the texture of the meat is more smooth and tender. The size of the lamb rack in this recipe can be cut into 9 pieces.*
- *Marinate the lamb rack with a little oil in order to get a light crispy texture.*
- *After mixing all marinade with the lamb rack, give each lamb piece a good massage to make sure they soak in the taste quickly and completely.*
- *Pan fry the lamb rack with hot oil over high heat helps keeping the juice locked inside the meat so that it won't get dry and tough.*

STIR FRIED AMARANTH WITH CHICKEN OIL

*Refer to p.87 for steps

INGREDIENTS
600 g amaranth
1 tbsp chopped garlic
chicken oil
1 red chili (shredded)

SEASONING
sugar
salt

FOR CHICKEN OIL
chicken fats
grated shallot

METHOD
1. Rinse chicken fats. Drain and leave aside for later use.
2. Heat up a wok and pour in a splash of oil. Put in the chicken fats and pan fry for a while. Add some water to let the oil of the chicken comes out faster. Add in grated shallot. When aroma of the shallot is released. Filter. Leave aside for later use.
3. Rinse and drain amaranth. Leave aside.
4. Heat up a wok and pour in oil. Put in the amaranth. Stir fry until 40% done. Add in garlic and sugar. Stir well. Stir in salt. Add in shredded red chili. Stir well. Serve. *Picture 1-3*

GRANDPA'S TIPS

- *Chicken oil is much flavorful than other kinds of oil such as peanut oil and corn oil.*
- *As garlic is very easy to getting burnt, add it into the wok when the amaranth reaches 40% done can prevent it gets burnt.*

STEAMED WINTER MELON WITH JINHUA HAM IN RECTANGLE SHAPE

Refer to p.91 for steps

INGREDIENTS

winter melon (the circle should be around 3 inches wide)
150 g Jinhua ham
1/2 egg white

THICK SAUCE

abalone sauce
sugar
caltrop starch solution
Gou Qi Zi

METHOD

1. Peel winter melon, make sure there's no more skin on the meat. Cut the winter melon into rectangle-shaped. Cut into pieces, the size of each piece should be 1 cm thick; 7.5 cm long and 5.5 cm wide. Slightly cut open the winter melon at the middle, vertically, but don't cut it through. *Picture 1-5*

2. Steam Jinhua ham for 45 minutes to an hour. Thinly slice to rectangle shape that is a bit smaller than the winter melon. The quantity of sliced Jinhua ham should be same as the winter melon. Stuff in one piece of Jinhua ham to one piece of winter melon. *Picture 6-7*

3. Steam the stuffed winter melon until it turns to transparent. It takes approximate 10 minutes, the time is depends on the thickness and the size of the winter melon.

4. Mix caltrop starch solution and abalone sauce well. Taste the abalone sauce before using, add a little sugar into it if it's too salty. Cook the mixture until thickens. Add in Gou Qi Zi and egg white. Push the mixture inside the wok gradually. Pour on the winter melon. Serve.

GRANDPA'S TIPS

- *Peel winter melon deep to make sure all the peel is gone as it has bitter taste.*
- *Since the quality of Jinhua ham could have big differences, taste before using. If you find the one you have is too salty, steam it for 45 minutes to an hour so the salty taste could be reduced.*
- *Want to serve this dish but want to do less? Try this simplified version: thinly slice winter melon and cook in hot water until soft, stuff in Jinhua ham and steam.*

STIR FRIED ASSORTED SEAFOODS WITH CHINESE CHIVES

Refer to p.95 for steps

INGREDIENTS
1 bundle Chinese chives
38 g dried shrimps
38 g dried whitebaits
1 dried squid
2 squids (small)
300 g fresh prawn meat
3 stalks preserved radish (the amount can be adjusted)
red bell pepper shreds

SEASONING
1/2 tsp salt
1 tsp sugar

METHOD
1. Soak dried squid in water until soft. Shred. Gut and rinse squids. Make some light cuts on the squids with a knife. Shred preserved radish. *Picture 1-4*

2. Cut off the old parts near the roots of the Chinese chives. Cut into sections. Divide the sections according to its position: top; middle and flower bulbs.

3. Soak dried shrimps in water until soft. Drain and blow dry. Heat up a wok and pour in oil. Deep fry dried shrimp until fragrant.

4. Soak dried whitebaits well. Drain. Blow dry. Heat up a wok and pour in oil. Deep fry the dried whitebaits until fragrant. Drizzle a splash of water to make the whitebaits crunchy and crispy. Keep deep frying the whitebaits until it turns light brown. Take it out from the wok. Darin off the oil.

5. Heat up a wok and pour in oil. Stir fry dried shrimps until fragrant. Add in whitebaits and stir well. Put in the dried squid and stir. Add squids and prawn meat. Stir fry until the prawn meat is almost cooked. Add sugar and stir well. Put in the shredded preserved radish and the top section of the Chinese chives. Stir fry until its colour changed. Put in the middle part of the Chinese chives and stir well. Add the flower bulbs and salt. Stir fry well. Serve. *Picture 5-6*

6. Add red bell pepper shreds for garnishment.

GRANDPA'S TIPS

- *Before setting the ingredients on the plate, you may add some deep fried pine nuts; peanuts or cashew.*
- *Add sugar to keep the green color of the Chinese chives.*
- *Why I divided the Chinese chives into 3 parts to cook? It's because the cooking time for those 3 parts are different.*
- *Don't add garlic, or the flavor of Chinese chives will be covered by it.*

STEWED POMELO PEEL IN TRADITIONAL WAY

Refer to p.99 for steps

INGREDIENTS

4 pc large thick pomelo peel
900 g dace fish bone
2 dried flatfish
75 g dried shrimps
lard
2 tbsp sugar
1 tsp salt
bamboo mat

GRANDPA'S TIPS

- *Let the pomelo peel soak in enough lard, then you can enjoy a velvet smooth and melt-in-your-mouth texture from it. How to know when the peel absorbed enough lard? Use a spoon to press the peel once and if there's no bubble appears which means it had soaked up enough lard.*

- *Homemade lard is very easy to make! Buy fatty pork meat from butcher. Rinse and drain. Put it into a wok and pour in some water. Boil over low heat until the fats come out.*

- *Make sure pomelo peel is burnt enough, or the bitter taste can't be removed.*

- *Soup made from dace fish is fresh and sweet in taste. If you use other fish, such as bighead carp, to cook soup, the soup will have fishy smell.*

- *Scoop out the bubbles during cooking fish soup, or the soup would taste bitter.*

METHOD

1. Roast pomelo peel over a fire until it gets burnt and turns to black color. Soak in water for about 7-8 hours; or overnight if possible; until the burnt skin begins to peeling off which means the soaking process is completed, then you may rub away the burnt skin. Rinse and squeeze well. Tear off the membrane of the pomelo peel. Cut the peel into oval shape. Put under the tube and rinse with running water for 2 days until its color turns to bright white, squeeze water out from the peel at least 7-8 times during rinsing to remove the bitterness. *Picture 1-6*

2. Pour in lard into a pot. Put in the pomelo peel. The amount of lard should be enough to cover the pomelo peel so that the pomelo peel can soak it up well.

3. Roast dried flatfish over a fire until fragrant. Scrape off the burnt parts and the head. Cut into large pieces. *Picture 7-10*

4. Bring a pot of water in a boil. Put in 2 tbsp of oil and 1 tsp of salt. Add dace fish bone, dried shrimps and dried flatfish into it. Boil over high heat for 30 minutes until the dried shrimps and dace fish bone release their flavors. Put in the pomelo peel and press bamboo mat on top to fix the peel's position so that it would not move during boiling. When the water is boiled, move the lid in a slanted position so that there's a gap between the lid and the pot. Turn to medium heat and boil for a while. Turn to low medium heat and boil for 3 hours. *Picture 11-12*

5. Put pomelo peel on to a plate. Bring a proper amount of the soup from the pot in a boil. Pour on top of the pomelo peel. Serve when it's hot. *Picture 13-14*

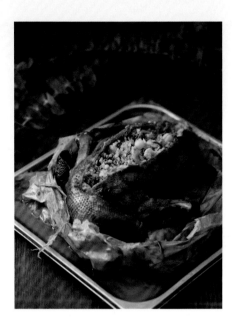

INGREDIENTS

1 duck (approx. 1.5 kg)
4 dried tangerine peel
38 g raw Job's tear
38 g rice
1/4 roasted duck
75 g fresh prawn meat
19 g dried black mushrooms
113 g fresh chestnuts
75 g fresh lotus seeds
38 g walnuts
38 g deep fried dried shredded scallop (see P. 76 for the method)
4 salted duck egg yolks
4 pc mulberry paper

COLORING

pearl soy sauce

DRIED TANGERINE PEEL EIGHT TREASURES DUCK

Refer to p.102-103 for steps

METHOD

1. Soak dried tangerine peel until soft. Scrape off the pitch. Cut 2 to 3 pieces into shreds.

2. Soak rice and Job's tear well. Drain. Remove the cores of the lotus seeds. Soak dried black mushrooms until soft. Cut roasted duck and black mushrooms into dices. Cut chestnuts and walnuts into halves. Cut each of the salted duck egg yolks into 4 pieces. Cut the membrane off from the prawn meat. Mix all of these ingredients with dried shredded tangerine peel as a filling.

3. Cut the gonads and the webs off from the duck. Debone the duck from its butt. Use a small knife to pull out the chest bone. Stuff in the filling until 60% full. Use metal skewer to stitch the duck. *Picture 1-10*

4. Pour in cold water to a wok. Blanch the duck without cover the lid, keep scooping boiling water on to the duck during blanching. Take out the duck. Leave until it cools a bit that won't burn your hands. Paint some pearl soy sauce on it. Basting stitch the duck with the metal skewer. *Picture 11-12*

5. Heat up oil. Put in the duck and blanch until the skin tightens. Keep pouring the boiling oil on to the duck with a soup ladle during blanching. Take the duck out from the wok and drain. *Picture 13-15*

6. Spread some tangerine peel on mulberry paper. Put the duck on top of it, then arrange the rest of the tangerine peel on it. Wrap. Put a piece of tangerine peel on the mulberry paper. Put it into a steamer and steam for 3 hours. Serve. *Picture 16-21*

STEAMED CHICKEN WITH MULBERRY LEAVES, MUSHROOMS AND CORDYCEPS

Refer to p.107 for steps

INGREDIENTS

1 game hen (approx 1.2 kg)
approx. 20-25 pc fresh mulberry leaves
19 g dried daylily
19 g black fungus
3 dried black mushrooms
19 g cordyceps

MARINADE

sugar
salt
caltrop starch
boiled oil

METHOD

1. Soak dried daylily, black fungus, dried black mushrooms and cordyceps separately, in water until soft. Cut the hard stalk off from the daylily. Remove the sawdust on the stem off from the black fungus. Cut the stalk off from the black mushrooms and shred it. *Picture 1-5*

2. Debone the game hen and thinly slice. Season the game hen with a pinch of sugar. Mix it well with daylily, black mushrooms, black fungus and cordyceps. Marinate with salt, caltrop starch and boiled oil for a while.

3. Rinse mulberry leaves and soak up the water with a dry towel. Remove the stems. *Picture 6*

4. Spread the mulberry leaves all over a plate. Place a piece of sliced game hen on each of the mulberry leaves. Then put the rest of the ingredients on top. *Picture 7-9*

5. Bring a pot of water to a boil over high heat. Put the plate into the pot and steam for 10 minutes. Turn off the heat and cover for a while. Serve. *Picture 10*

GRANDPA'S TIPS

- *Mulberry leaves can clear Heat and moisten the lungs; loosen the bowel to relieve constipation; and helps eliminate toxins from the body. Before steaming, it tastes bitter with rough texture. Thanks to the moisturizing effect of the chicken oil, the texture of mulberry leaves becomes smooth after steaming with it. I suggest using the fresh and young one.*

- *Mulberry leaves could be picked up and had been kept for a while before you buy it. Soak it into cold water can keep it fresh and straight.*

- *Slice the game hen as thin as you can as the mulberry leaves can't be cooked for too long.*

- *The reason to cover the chicken and leave for a while after it's cooked, is to allow it has time to soak in the sauce so that you can enjoy a juicy and tasty chicken.*

- *Suggest using small black fungus, it is easy to cook and crispy texture.*

- *The steaming time for this dish could be adjusted according to the thickness and size of the chicken.*

POACHED CHICKEN IN SOY SAUCE WITH ROSE WINE

Refer to p.111 for steps

GRANDPA'S TIPS

- *Don't skip step 1 as it's very important! The reason to do this is to tighten the skin of the chicken so that the chicken would not break while poaching.*

- *For chicken that weight only 1.2 kg, soak for 20 minutes after the heat is turning off would be fine enough.*

- *The reason to add rock sugar is to make chicken meat more silky.*

- *Rose wine should be added into the pot at the last stage of the cooking, if you pour in it too early, the wine flavor would be evaporated.*

- *To test the chicken and see if it is cooked, prick the meat in-between the wings and thighs, if the chopstick could prick through easily means it's cooked. If it's undercook, just put it back to the marinade sauce and poach a bit longer.*

- *Wait until the chicken cool-down, then cut it to prevent the skin pulls back or the skin and bone fall apart.*

- *Store the marinade in the fridge for next time. It can be used for poaching chicken thigh, chicken wing or even pork belly.*

INGREDIENTS
1 chicken (approx. 1.8 kg)
75 g rock sugar
1 cup dark soy sauce
1/2 cup rose wine
straw strings

MARINADE SAUCE
1 dried tangerine peel
3 cloves star anise
4 cinnamon
8 bay leaves
5 licorice
3 cloves Cao Guo

DIPPING SAUCE
(add a little salt into each and pour in boiling oil)
chopped ginger
chopped spring onion with ginger
chopped shallot
chopped shallot with ginger

METHOD

1. Rinse chicken. Tie the chicken wings together tightly with straw strings. Hold it and dip the chicken into boiling water. Meanwhile, pour the boiling water inside the chicken. *Picture 1-2*

2. Pour in water to a pot and the amount should be enough to cover the chicken. Put it all marinade sauce ingredients and bring to a boil. Then keep boiling until the flavor comes out. Add in rock sugar and boil until it is melted. Add dark soy sauce and chicken with the back facing down. Boil for 8-10 minutes. Cover with a lid. Turn off the heat. Leave for 25 minutes. Add in rose wine and leave for 5 minutes. *Picture 3-5*

3. Take the chicken out from the pot. Hang and air dry it for 20 minutes. Cut the chicken into pieces after cool-down. Serve with dipping sauce. *Picture 6-9*

SAGO SWEET SOUP WITH GE XIAN MI

INGREDIENTS

10 g Ge Xian Mi
1/2 measuring
cup sago
150 g rock sugar
2-3 sliced ginger

METHOD

1. Soak Ge Xian Mi in water for an hour until it expands in size. Put it into a pot of boiling water with sliced ginger. Scald and drain well. Rinse with cold water. Drain well. (approximate quantity of the rice is 1 bowl).

2. Put sago into a pot of boiling water and boil for 10 minutes. Cover with a lid and leave for 10 minutes. Rinse with cold water and drain well.

3. Boil a proper quantity of water until it's boiling. Add rock sugar and boil until it dissolves (taste before going to the next step). Put Ge Xian Mi into the pot and bring to a boil, keep boil for a while. Add sago. Bring to a boil again. Serve.

GRANDPA'S TIPS

- Ge Xian Mi is a wild nostoc that grows in freshwater, and the water must be crystal clear and pollution-free. After soaking, the Ge Xian Mi became beautiful round shape with dark green color which looks like green pearl.

- Sago Sweet Soup with Ge Xian Mi was popular during the 60s and 70s in the last century. Men used to celebrate their 60's birthday with dessert made from Ge Xian Mi; and women would use Longevity Peach Lotus Paste Bun instead. The implied meaning of these two birthday desserts are different.

- As a member of the Nostocaceae family, Ge Xian Mi has a mild unpleasant smell. Therefore, scald it in boiling water with sliced ginger to remove the smell.

- Better use the Ge Xian Mi immediately after soaking. If you leave it in the refrigerator and it will soak in moisture from the fridge, and the texture will become very soft and sticky while cooking.

- To make a perfect round shaped sago, don't soak it in water! Instead, put it into boiling water directly and stir well, cook for a while and cover until the color turns transparent. Then, rinse with cold water.

- Can add coconut milk or almond juice into the sweet soup as you like.

Xian Wen Mi, another name of the Ge Xian Mi, in Chinese means a man who lives for a long time. Because of this blissful meaning man who reaches his 60th birthday will take this sweet soup for celebration.

DRIED FISH MAW SWEET SOUP WITH RED DATES

*Refer to p.116 for steps

INGREDIENTS

75 g sweet apricot kernel
(The proportion of sweet and bitter apricot kernel is 5 to 1)
15 g bitter apricot kernel
3 dried fish maws (weight: 40 units per 1 catty)
1 pc dried tangerine peel
8 red dates
190 g rock sugar
3 sliced ginger (with skin on)
2 sprigs spring onion

METHOD

1. Soak dried fish maws for 2 hours. Put ginger, with skin on, and spring onion into a cold water. Bring to a boil. Put in the fish maws. Turn off the heat. Cover and leave for 2-3 hours until the fish maw is tender. Take it out from the water. Cut into pieces. *Picture 1-3*

2. Soak dried tangerine peel until soft. Scrape off the pith.

3. Rinse sweet and bitter apricot kernel. Rinse and core red dates.

4. Bring a proper quantity of water in a boil. Put in the apricot kernel and boil for 45 minutes until tender. Add rock sugar and boil until it is dissolved. Put dried tangerine peel and red dates into it. Boil for 20-30 minutes. Taste. Turn to high heat. Put in the fish maws. Bring to a boil again. Cover with a lid and leave for 5 minutes. Serve. *Picture 4-6*

GRANDPA'S TIPS

- *Add dried fish maw at the last stage of the cooking, or it will be dissolved completely.*
- *The proportion of sweet and bitter apricot kernel is 5 to 1. It's because bitter apricot kernel contains a trace of toxin that you should not take too much.*

SESAME SWEET SOUP WITH FINELY SHREDDED BEAN CURD

*Refer to p.118-119 for steps

INGREDIENTS

300 g black sesame
white sesame
rock sugar
1 bowl peanuts (lightly salted deep fried peanuts)
1 bowl rice
water chestnuts powder mixture
1 box silky smooth bean curd (finely shredded)

METHOD

1. Stir fry black and white sesame without oil until fragrant. Leave to cool. *Picture 1-2*

2. Stir fry rice without oil until fragrant.

3. Put sesame, peanuts and rice into a blender. Pour in water. Blend until mushy. Drain with a muslin bag.

4. Bring sesame paste in a boil. Pour in water chestnuts powder mixture. Stir well. Put in the finely shredded bean curd. Look, how muslin it is!

Follow my way to shred the bean curd (Vance Tofu) on p.119

GRANDPA'S TIPS

- *The proportion of black sesame, stir fried rice, deep fried peanuts to water is 1 to 1.5.*
- *People who has peanut allergy can't take this sweet soup.*

GREENLIP ABALONE AND SILKIE CHICKEN SOUP WITH ASSORTED MUSHROOMS

Refer to p.124 for steps

INGREDIENTS

1 greenlip abalone
19 g dried blaze mushrooms
19 g yellow morel
19 g dried cordyceps flower
1 silkie chicken
38 g Jinhua ham
6 chicken feet
300 g lean pork
coarse salt

METHOD

1. Rinse silkie chicken. Cut into large pieces. Rinse lean pork. Cut into large pieces. Cut Jinhua ham into large dices.

2. Pour cold water into a pot. Put in chicken feet, silkie chicken, lean pork and Jinhua ham. Blanch and rinse. Cut the nails off from the silkie chicken and the chicken feet. Make a light cut on the thickest part of the soles so that the collagen could be released completely during boiling. *Picture 1-3*

3. Soak blaze mushrooms, yellow morel and cordyceps flower until soft. Rinse. Cut the stems off from the blaze mushrooms and yellow morel. Leave the mushrooms and cordyceps flower aside for later use. *Picture 4-5*

4. Cut off the mouth of greenlip abalone. Wash with coarse salt. Rinse. Slice at an angle so that its flavor could be released more easily while boiling.

5. Put lean pork, chicken feet, some abalone slices, silkie chicken, yellow morel, blaze mushrooms, cordyceps flower and Jinhua ham into a slow cooker. Arrange the rest of the abalone slices on top. Pour in boiling water. Seal the slow cooker with microwave wrap and mulberry paper. Stew for 3 hours. Serve. *Picture 4-5*

GRANDPA'S TIPS

- *Greenlip abalone from South Australia is big and meaty, therefore, the soup would be sweet enough with one piece of it. For local abalone, though it is fit for soup as well, its taste is not as good as the one from South Australia. If you want to use the local one, should take 2 pieces.*
- *Since the stems of yellow morel has bitter taste, better remove it before using the morel.*
- *Slice abalone at angles is to let it releases flavor more easily during boiling the soup. Picture 6 Thinly slice the abalone if it is used for steaming. Picture 7-8*

FISH THICK SOUP

Refer to p.127 for steps

Refer to p.127 for steps

GRANDPA'S TIPS

- *This dish is a dish original from Shunde, a place which is a home of fish and rice; and is famous for fresh fish cuisine. Though the taste of river food is not as sweet and fresh as seafood, the dish made from river food could be tasty if you know how to make it.*

- *Debone the bonito before cooking it helps the flavor comes out faster. Bonito is the "spirit" of the soup that it enhances the sweet and fresh flavors.*

- *Finely shred the lettuce's stems and you can enjoy a crispy and crunchy mouthfeel.*

INGREDIENTS

1 bighead carp (approx. 600 g)
1 large wood ear (shredded)
1 carrot (shredded)
2 fresh bamboo shoot (shredded)
1/2 angled luffa (shredded)
1 sea cucumber (shredded)
shredded ginger
2 stalks Chinese lettuce
water chestnuts powder mixture
salt

SOUP BASE

1 white crucian carp
6 dace fish bones
2 bonito
sliced ginger

METHOD

1. Soup base: pick out the meat of bonito. Pan fry white crucian carp and dace fish bones with sliced ginger until fragrant. Pour in water. Add bonito and heat up. Boil until it becomes milky fish soup. Strain the soup. Keep the bonito. Make sure all bones are gone. Leave the meat aside for later use.

2. Cut the belly off from the bighead carp. Cut open the head and cut into pieces. *Picture 1-4* Put the head on a steaming plate. Spread shredded ginger on top. Steam for 5 minutes until it reaches 60% done. Put in the fish belly and arrange shredded ginger on top. Continue to steam for 8 minutes.

3. Leave the fish head and fish belly to cool. Remove the bones and keep the meat for later use.

4. Finely shred the stems of the lettuce. *Picture 5-7*

5. Heat up the soup base. Put in the fish meat, sea cucumber, wood ear, carrot, bamboo shoot and angled luffa. Season with a little salt. Bring to a boil again. Pour in the water chestnuts powder mixture to thicken the soup. Add in the shredded lettuce. Serve when it's hot. *Picture 8-9*

STEWED CHICKEN SOUP WITH GINKGO AND DRIED BEAN CURD SHEET

INGREDIENTS

1 chicken
300 g lean pork
Jinhua ham
113 g ginkgo
1 dried bean curd sheet
3 dried bean curd sticks

METHOD

1. Cut lean pork and chicken into large pieces, separately. Remove the nails from the chicken feet. Scald chicken with lean pork and Jinhua ham. Rinse well.

2. Shell ginkgo and put it into boiling water. Take it out and rinse with cold water when its skin begins to peeling off. Rub it gently to remove the skin.

3. Soak dried bean curd sheet and sticks in water until tender. Cut the sheet into large pieces and cut the sticks to 3 inches long.

4. Put dried bean curd sticks, chicken and lean pork into a slow cooker. Wrap around the side with the dried bean curd sheet. Arrange ginkgo and Jinhua ham on top. Pour in boiling water till the slow cooker is 90% full. Cover with a lid. Seal with microwave wrap or mulberry paper. Stew for about 3 1/2 hours. Serve.

GRANDPA'S TIPS

- *You may wonder why I didn't remove the chicken skin. The reason is the chicken skin can make the soup fragrant. If you don't like the fats, just remove as much as you can before using the chicken .*
- *The purpose of adding lean pork and Jinhau ham into the soup is to enrich the flavor. The taste of the soup would rather blend If it's only cooked with chicken, bean curd sheet and ginkgo.*

CORDYCEPS SOUP WITH CHESTNUT AND CASHEW

INGREDIENTS
300 g cashew
113 g lotus seeds (Xiang Lianzi)
150 g peanuts
900 g fresh chestnut (peeled)
a few cordyceps
19 g Gou Qi Zi

METHOD
1. Soak cashew; lotus seeds and peanuts in water, separately, for 2 hours and drain. Rinse cordyceps and Gou Qi Zi.

2. Put chestnut into boiling water, boil until the water is boiling again. Thus, the skin of the chestnut will be fallen apart. Take out the chestnut and leave it aside for later use.

3. Apart from Gou Qi Zi, put all ingredients into a pot. Pour a proper quantity of water to cover the ingredients. Boil over high heat until it's boiling. Turn to low heat and boil until the peanuts soften. Then add Gou Qi Zi into the pot. Cook for 1 1/2 hours approximately. Boil another 30 minutes and serve.

4. Taste first before serving. Let guests decide how many salt they want to put into their bowl while taking the soup.

GRANDPA'S TIPS

- *This soup can cleanse the gastrointestinal tract and improve kidneys function.*
- *Choose cordyceps with lightweight and dry texture.*

JI GU CAO, XI HUANG CAO AND JIN QIAN CAO DRINK

INGREDIENTS
150 g Ji Gu Cao
150 g Xi Huang Cao
150 g Jin Qian Cao
8 candied dates

METHOD
1. Rinse Ji Gu Cao, Xi Huang Cao and Jin Qian Cao, soak them into cold water overnight.

2. Next day, drain the water out from the herbs. Pour another pot of water, put Ji Gu Cao, Xi Huang Cao and Jin Qian Cao into it. Scald.

3. Take the Ji Gu Cao, Xi Huang Cao and Jin Qian Cao out. Put them into another pot. Add candied dates and a proper quantity of water in. Boil until the water is boiling, then boil for 3 or 4 hours. Serve.

GRANDPA'S TIPS

- *Ji Gu Cao, Xi Huang Cao and Jin Qian Cao are all dry foods that you can buy at Chinese medicine shop.*
- *Soak Ji Gu Cao, Xi Huang Cao and Jin Zian Cao overnight not only can remove the sands from their leaves, but also release their flavors.*
- *This drink can clear Heat and eliminate toxins from the body. It's a drink for all seasons.*
- *This recipe serves 6 to 8 persons.*

STIR FRIED TARO RICE WITH PORK CRACKLING

INGREDIENTS

113 g taro
300 g fatty pork meat
2 bowls cooked rice
38 g dried shrimps
shallot

SEASONING

sugar
fish sauce

METHOD

1. Soak dried shrimps in water until soft. Finely chop and leave aside for later use.

2. Cut taro into square dices. Leave aside for later use.

3. Dice fatty pork meat. Leave aside for later use.

4. Heat wok with oil. Put in the diced fatty pork meat. Deep fry the fatty pork until the oil begins to come out, add a splash of water to help pork releases oil. When the pork is crispy and fragrant, take it out. Drain off the oil. Leave aside.

5. Deep fry taro with pork fat until cooked and crispy. Take out from the wok. Drain off the oil. Leave aside for later use.

6. Heat wok with pork fat. Lower the heat a bit. Add in some shallot and stir fry until fragrant. Put the cooked rice into it and stir well. Add dried shrimps. Stir fry. Put in the taro and season with a little sugar. Put in the remaining shallot and pour in fish sauce. Stir fry well. Put in the pork crackling. Serve when it's hot.

GRANDPA'S TIPS

- *Buy pork fat which is from the back as fat from this part will be super crispy and flavorsome after deep frying.*
- *Blanch taro in oil not only can get a crispy mouthfeel, but also prevent the taro from breaking or loosed when stir frying.*

CONGEE WITH HOT AND SPICY DRIED SEAFOOD ON THE SIDE

Refer to p.136 for steps

GRANDPA'S TIPS

- *A plain white congee with a dish of delicious stir-fried dried shrimp and dried fish with fermented black beans on the side may increase your appetite during the hot season, especially when you have no motivation to eat.*

- *It's not necessary to wash the dried whitebaits, instead, remove the dirt inside would be fine enough. As dried whitebaits will become tender after washing; or you have to air-dry it if it's been washed.*

- *I will add a splash of water into the dried whitebaits during stir-frying in order to make it more crispy and crunchy. However, it's quite danger as the boiling oil will spill out that you must be very careful. Picture 6-8 Using too much water will let the fish "explodes" like what you see in the picture. Picture 9*

- *I like using new rice to make congee as it has plenty of moisture and the texture is soft that is easy to becomes smooth and silky.*

- *You can adjust the quantity of the ingredients to fit your own taste.*

INGREDIENTS

75 g dried prawn
38 g dried whitebaits
1/2 clove garlic (chopped)
4 cloves shallot (chopped)
2 tbsp fermented black beans
(chopped briefly)
2 red chilies (diced)
sugar
salt
dark soy sauce (for coloring)

CONGEE'S INGREDIENTS

new rice
1 dried bean curd sheet

METHOD

1. Rinse and drain rice. Put it into a pot of boiling water. Cook until the rice is smooth and silky. Put in the dried bean curd sheet and cook until it's melted.

2. Deep fry dried whitebaits with oil until crispy and crunchy. Leave aside for later use. *Picture 1-4*

3. Soak dried prawn in water until it softens. Take it out and drain well. Stir fry briefly with oil.

4. Heat up a wok. Pour in oil. Put dried prawn, chopped garlic, chopped shallot, whitebaits and fermented black beans into the wok, one by one in order. Stir well. Add in diced red chili. Season with sugar. Stir fry well. Add salt and stir fry again until all ingredients soak in the taste completely and evenly. Pour a splash of dark soy sauce for coloring. Serve.

DEEP FRIED SHREDDED WHITE RADISH CAKE

*Refer to p.139 for steps

INGREDIENTS
300 g white radish
1 carrot
38 g dried shrimps
113 g plain flour
19 g rice flour
1/2 tsp baking powder
2 sprigs spring onion (diced)

SEASONING
ground pepper
1/2 tsp five spice powder
1/2 tsp salt
1 tsp sesame oil

METHOD
1. Grate white radish and carrot into thin shreds. Squeeze out the excessive moisture of the white radish with our hands.
2. Finely dice dried shrimps. Sieve plain flour, rice flour and baking powder well.
3. Mix the white radish, carrot and the flour mixture well. Add in dried shrimps, five spice powder and ground pepper. Mix well. Pour in water. Stir well and add in sesame oil. Stir well. Put in spring onion. *Picture 1-4*
4. Heat up a wok and pour in oil. Put the round white radish cake mould into the wok to let it soak in to oil. Pour out all the oil. Fill in the white radish batter until 80% full. Put back to the wok and deep fry until the dough falls out from the mould and turns to golden brown. Dish up and drain. *Picture 5-7*

GRANDPA'S TIPS

- *It's not necessary to shred white radish fine. Just follow the sample in the picture for the thickness. If you the shreds is too thin will release water.*
- *Should put the mould into the oil for soaking each time before you fill in the batter so that the white radish cake could be fallen from it more easily.*

把握良機學習正宗法國廚藝

全港最具規模的煤氣烹飪中心聯同雲集全球頂尖名廚的Disciples Escoffier法國廚師會，為香港帶來首個具完整架構的傳統法國美饌烹飪課程。

法國烹飪藝術文憑

法國糕餅藝術文憑

5大優勢

☆ 學員通過考試可獲頒發「法國廚藝訓練第五級證書」，取得在法國開設餐廳的資格

☆ 由米芝蓮星級或國際知名大廚親授烹飪秘技

☆ 毋須抽身離港上課

☆ 在全新極專業、設備極完善之煤氣烹飪中心上課

☆ 由專業導師提供針對學員個人及全面性的培訓

報名及查詢請致電
2576 1535 / 6387 3157

明火煮食，鑊氣十足，煤氣煮食爐火力特強，
能均勻加熱整個鍋身，防止營養流失，
讓您輕易炮製不同美味佳餚，一同享受
明火煮食的樂趣！

明火煮食 樂趣多

超強火力

TGC 極炎火嵌入式平面爐
MEGA2

🔥 6.0 千瓦爐頭，火力特強

🕐 預校熄火時間功能，煮食更簡易

▨ 德國頂級陶瓷玻璃

時尚色彩

TGC 密封爐頭嵌入式平面爐
TRTB62ST-G

🎨 多種顏色選擇，配合時尚家居

🔲 密封式爐頭設計，方便清潔

🔄 火力達 5.0 千瓦，兼備獨立芯火

煮飯必備

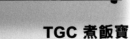

TGC 煮飯寶
RJ3R

🍚 內置煮飯功能

🔥 6.6 千瓦特大火力炒鑊爐頭

鼎爺廚房
原汁原味功夫菜
3

作者	Author
李家鼎	Steve Lee

策劃/編輯	Project Editor
譚麗琴	Catherine Tam
簡詠怡	Karen Kan

攝影	Photographer
	Imagine Union

美術統籌及設計	Art Direction & Design
羅美齡	Amelia Loh

美術設計	Design
鍾啟善	Nora Chung

出版者 Publisher
萬里機構出版有限公司 Wan Li Book Company Limited
香港鰂魚涌英皇道1065號 Room 1305, Eastern Centre, 1065 King's Road,
東達中心1305室 Quarry Bay, Hong Kong.
電話 Tel: 2564 7511
傳真 Fax: 2565 5539
電郵 Email: info@wanlibk.com
網址 Web Site: http://www.wanlibk.com
　　　　http://www.facebook.com/wanlibk

發行者 Distributor
香港聯合書刊物流有限公司 SUP Publishing Logistics (HK) Ltd.
香港新界大埔汀麗路36號 3/F., C&C Building, 36 Ting Lai Road,
中華商務印刷大廈3字樓 Tai Po, N.T., Hong Kong
電話 Tel: 2150 2100
傳真 Fax: 2407 3062
電郵 Email: info@suplogistics.com.hk

承印者 Printer
中華商務彩色印刷有限公司 C & C Offset Printing Co., Ltd.

出版日期 Publishing Date
二零一九年七月第一次印刷 First print in July 2019